Regularity of Difference Equations on Banach Spaces

Ravi P. Agarwal • Claudio Cuevas • Carlos Lizama

Regularity of Difference Equations on Banach Spaces

 Springer

Ravi P. Agarwal
Department of Mathematics
Texas A&M University
Kingsville, TX, USA

Claudio Cuevas
Departamento de Matemática
Universidade Federal de Pernambuco
Recife, PE, Brazil

Carlos Lizama
Departamento de Matemática y
Ciencia de la Computación
Universidad de Santiago de Chile
Santiago, Chile

ISBN 978-3-319-35518-4   ISBN 978-3-319-06447-5 (eBook)
DOI 10.1007/978-3-319-06447-5
Springer Cham Heidelberg New York Dordrecht London

Mathematics Subject Classification: 39A12, 39A06, 39A60, 47D09

© Springer International Publishing Switzerland 2014
Softcover reprint of the hardcover 1st edition 2014

Printed on acid-free paper

Springer is part of Springer Science+Business Media (www.springer.com)

Preface

Evolutionary equations in a Banach space X of the form

$$u(k + 1) - (\mathcal{A}u)(k) = f(k), \quad k \in \mathbb{Z}_+, \qquad (*)$$

arise in several branches of science and technology [1,26,61,75,81,89,116,125,129, 144,146]. Depending on the properties of the function $\mathcal{A} : \mathcal{F}(\mathbb{Z}_+; X) \rightarrow \mathcal{F}(\mathbb{Z}_+; X)$ where $\mathcal{F}(\mathbb{Z}_+; X)$ is a space of vector-valued sequences, such equations exhibit many new and interesting phenomena. Here, many problems still need to be solved, in particular in connection with nonlinear analogues of $(*)$. For example, there are only few results dealing with the semilinear case when the right-hand-side function depends not only on the discrete time k but also on the unknown solution u, i.e., $f = F(k, u)$.

Suppose that we know something about the behavior of the forcing function f in $(*)$. For example, f could be bounded, asymptotic in some sense, or f might satisfy $f \in l_p(\mathbb{Z}_+; X)$, where X is a Banach space and $1 \leq p \leq \infty$. The *maximal regularity* problem is then to find conditions on the data $\mathcal{A}$ so that the solution u of $(*)$ has the same behavior as f.

Blunck considered in 2001 [22, 23] the maximal regularity problem for the discrete-time evolution equation

$$x(k + 1) - Tx(k) = f(k), \quad k \in \mathbb{Z}_+,$$

on the vector-valued sequence spaces $l_p(\mathbb{Z}_+; X)$, i.e.,

$$\mathcal{A} : l_p(\mathbb{Z}_+; X) \rightarrow l_p(\mathbb{Z}_+; X)$$

is defined by $(\mathcal{A}u)(k) = Tu(k)$, where T is a bounded linear operator defined on a Banach space X which belongs to the class $\mathcal{H}\mathcal{T}$, that is, the space X satisfies the property that the Hilbert transform defined by

$$(Hf)(t) := \lim_{\substack{\epsilon \to 0 \\ R \to \infty}} \frac{1}{\pi} \int_{\epsilon < |s| < R} \frac{f(t-s)}{s} ds \quad (**)$$

is bounded in the vector-valued Lebesgue space $L^p(\mathbb{R}; X)$ for some $p \in (1, \infty)$. The limit in $(**)$ is to be understood in the L^p-sense. Such Banach spaces are also called *UMD*, by *unconditional martingale difference*, because they have an important role in probability theory [65].

In their work, Blunck has characterized the discrete maximal regularity of the above first-order evolution equation by two types of conditions: firstly, by R-boundedness properties of the discrete-time semigroup $(T^k)_{k \in \mathbb{Z}_+}$ and of the resolvent operator $R(\lambda, T) := (\lambda - T)^{-1}$. We recall from [65] that a subset $\mathcal{T}$ of the set of all bounded operators defined on X is called R-bounded (or Rademacher-bounded), if there exists a constant $C > 0$ such that

$$\|(Tx_1, \ldots, Tx_n)\|_R \le C \|(x_1, \ldots, x_n)\|_R,$$

for all $x_1, \ldots, x_n \in X$, $n \in \mathbb{N}$, and all $T \in \mathcal{T}$, where

$$\|(x_1, \ldots, x_n)\|_R := \frac{1}{2^n} \sum_{\epsilon \in \{-1,1\}^{\mathbb{N}}} \| \sum_{j=1}^{n} \epsilon_j x_j \|.$$

Secondly, by the maximal regularity for the continuous-time evolution equation, i.e.,

$$u'(t) - Tu(t) = f(t), \quad t \ge 0.$$

See also Portal [160, 162].

In the continuous case, it is well known that the study of maximal regularity is very useful for treating semilinear and quasilinear problems. Fourier multiplier theorems are among the most important tools to prove maximal regularity. They play an important role in the analysis of elliptic and parabolic problems. In past years it has become evident that one needs not only the classical theorems but also vector-valued extensions with *operator-valued* multiplier functions or symbols. These extensions allow one to treat certain problems for evolution equations with partial differential operators in an elegant and efficient manner in analogy to ordinary differential equations (see, e.g., Amann [6], Denk et al. [65], Clément et al. [46], the survey by Arendt [10], and the bibliography therein).

However, we note that for nonlinear discrete-time evolution equations, some additional difficulties appear. In fact, we observe that this approach cannot be done by a direct translation of the proofs from the continuous-time setting to the discrete-time setting. Indeed, the former only allows to construct a solution on a possibly very short time interval, the global solution being then obtained by extension results. This technique will obviously fail in the discrete-time setting, where no such thing as an arbitrary short time interval exists.

In this book we propose a way around the *short time interval* problem by assuming a summability in time on the constants appearing in Lipschitz-type conditions for the nonlinearity. This allows us to run a natural argument: Assuming maximal regularity of the linear part to obtain a priori estimates, we use these estimates together with adequate assumption on the nonlinearity to obtain a solution as a fixed point of suitable operator defined in a suitable discrete Sobolev-type space. We emphasize that the implementation of this approach is a priori, not trivial, as the reader will perceive through this book.

From an applied perspective, we give a set of results and techniques on abstract difference equations which are interesting when applied to concrete difference equations.

This monograph is an outgrowth of the authors' research on the subject during the past 10 years. Our expectation is that this book will be very useful as a source of information and motivation to researchers who are entering the subject. We hope that the book is also valuable for mathematicians in related fields who are interested in new methods in Banach space theory for the purpose of using it in applied areas and contribute to the development of the theory.

We have tried to preserve an informal style and reduce technicalities as much as possible. We do include indications of proof of some results with the hope that this will help the reader to understand and get a feeling for the interplay of the various concepts and techniques which are discussed.

This book is divided into seven chapters. In the first chapter we consider several examples of discrete evolution equations arising in diverse contexts. The chapter also deals with some results on the transform method, which are used in applications (Sect. 1.2). We present the basic discrete semigroup theory and introduce the discrete cosine and sine operators (Sect. 1.4).

The second chapter consists of a collection of results that are scattered in many publications, essentially dealing with maximal regularity. In Sect. 2.1 we introduce the concept of UMD space and review its basic properties. In Sect. 2.2 we introduce the notion of R-boundedness, which has proved to be a significant tool to deal with maximal regularity. R-boundedness has a number of nice permanence properties which are summarized in this subsection. In Sect. 2.3 we give an overview of the main results on this subject. We note that one way to treat the problem of maximal regularity is by applying the operator-valued Fourier multiplier theorems. In Sect. 2.4 we collect some of the most important vector-valued multiplier theorems. In the discrete-time setting, the operator-valued Fourier multiplier theorem is due to Blunck (see Theorem 2.4.9).

Chapter 3 deals with Blunck's characterization of maximal l_p-regularity for first-order linear difference equations by R-boundedness properties of the semigroup associated. It is the indispensable basis for the theory of semilinear discrete evolution equations, which we present in Chap. 4. More precisely, it is devoted to the existence of bounded solutions for a first-order semilinear difference equation, whose first discrete derivative of the solution belongs to the Banach space of all p-summable sequences, denoted by l_p.

In Chap. 5 we introduce the notion of maximal l_p-regularity for second-order linear difference equations and we characterize the maximal l_p-regularity for such equations. We also include in this study results on exact discretizations of the harmonic oscillator (see Sects. 5.2, 5.3, and 5.4). The classical reference is Agarwal's book [1], where exact discretizations are discussed in detail, in addition to many applications. Section 5.5 deals with the problem of regularity in weighted spaces. In Sect. 5.6, the relation between well-posedness and the maximal regularity is presented.

Chapter 6 is concerned with the study of the existence of bounded solutions whose second discrete derivative is in l_p for second-order semilinear difference equations. The essential technique employed in this treatment is the knowledge of maximal regularity of the associated homogeneous equation. We also develop this theory for exact semilinear second-order equations and semilinear problems on weighted spaces. Questions regarding local perturbations have also been considered.

Chapter 7 has practical importance because it deals with applications of the theory of discrete maximal regularity to stability of concrete discrete process. We present a useful R-boundedness criterion and study both the boundedness and the asymptotic profile of the solutions of some discrete evolution equations. We use the methods presented in Chap. 4 to study the existence and uniqueness of bounded solutions which are in l_p for semilinear functional difference equations with infinite delay (RFDE for short). We also present results about asymptotic behavior for RFDE. In the last section of Chap. 7, we present applications to discrete Volterra difference equations with infinite delay.

We have included a section of comments at the end of each chapter, which describes further aspects of the theory. The book concludes with an extensive bibliography.

The authors acknowledge partially support from the following institutions and grants: C. Cuevas and C. Lizama by project Anillo ACT-1112 (CONICYT-Chile); C. Cuevas by Programa Atracción e Inserción (PAI-MEC) Grant 80112008 (CONICYT-Chile).

Parts of this book were written while the second author was visiting University of Nantes (September–November 2011) and University of La Frontera (July–September 2012). He wants to express his gratitude for the hospitality of the Departments of Mathematics at these institutions. Finally, the authors take this opportunity to express the appreciation and thanks to their respective wives: Sadhna Agarwal, Gilca Cuevas, and Maricel Lizama for the encouragement and understanding in writing this monograph.

Kingsville, TX   Ravi P. Agarwal
Recife, Brazil   Claudio Cuevas
Santiago, Chile   Carlos Lizama

Contents

List of Abbreviations

$X, Y \equiv$ Banach spaces.

$X \oplus Y \equiv$ The direct sum of X and Y.

$\mathbb{N} \equiv$ The set of natural numbers.

$\mathbb{N}(n_0) \equiv$ The set $\{n \in \mathbb{N}/n \geq n_0\}$.

$\mathbb{Z}_+ \equiv$ The set of positive integers including zero.

$\mathbb{R} \equiv$ The set of real numbers.

$\mathbb{R}^+ \equiv [0, \infty)$.

$\mathbb{C} \equiv$ The set of complex numbers.

$\mathbb{C}^r \equiv$ The r-dimensional complex Euclidean space.

$\mathbb{T} \equiv$ The unit circle in the complex plane.

$\mathbb{D} \equiv$ The unit disk in $\mathbb{C}$.

$\mathbb{D}(z, r) \equiv$ Denotes the set $\{w \in \mathbb{C} : |w - z| < r\}$.

$\mathbb{T}_r^\alpha \equiv$ Denotes the set $\{z \in \mathbb{C} : |z| = \alpha r\}$.

$\delta[n] \equiv$ Denotes the Kronecker's delta.

$\Delta_r \equiv$ Denotes the forward r-difference operator of the first order, i.e., for each
  $x : \mathbb{Z}_+ \to X$, and $n \in \mathbb{Z}_+$, $\Delta_r x_n = x_{n+1} - r x_n$, $r \in \mathbb{R}_+$. In the border case
  $r = 1$ we denote $\Delta \equiv \Delta_1$. Moreover, we denote $\Delta_r^2 x_n = \Delta_r(\Delta_r x_n)$.

$\mathcal{B}(X, Y) \equiv$ The space of bounded linear operators from X into Y endowed with the
  uniform operator topology. We abbreviate $\mathcal{B}(X)$ whenever $X = Y$.

Let $T \in \mathcal{B}(X)$. We denote by $||T||$ the norm of operator T.

Let $T \in \mathcal{B}(X)$. T^n stands for the nth iteration of T for each nonnegative integer n.

$M(T) \equiv \sup_{n \in \mathbb{Z}_+} ||T^n||$.

$R(\mathcal{T}) \equiv$ The R-bound of the set $\mathcal{T}$ (Sect. 3.2).

$L^p(a, b; X) \equiv$ The L^p-space of all functions X-valued and p-integrable on $[a, b]$.

$L^p(\mathbb{R}^n; X) \equiv$ The L^p-space of all functions X-valued and p-integrable on $\mathbb{R}^n$.

$l^p(\mathbb{Z}^+; \mathbb{C}^r) \equiv \{\varphi : \mathbb{Z}^+ \to \mathbb{C}^r \,/\, \|\varphi\|_p^p := \sum_{n=0}^{\infty} |\varphi(n)|^p < \infty\}$, $1 \leq p \leq \infty$ (see
  Sect. 7.7).

$l_p(\mathbb{Z}_+; X) \equiv$ The collection of all X-valued sequences so that

$$\|x\|_p := \left(\sum_{n=1}^{\infty} \|x_n\|^p \right)^{1/p}, \ 1 \le p < \infty.$$

$l_\infty(\mathbb{Z}_+; X) \equiv$ The collection of bounded X-valued sequences.

$l_p^r(\mathbb{Z}_+; X) \equiv \left\{ (x_n) : (r^{-n} x_n) \in l_p(\mathbb{Z}_+; X) \right\} \ (r > 0), \ 1 < p < +\infty$ (Sect. 5.5).

$l_{p,r}^1(\mathbb{Z}_+; X) \equiv \left\{ (y_n) : y_0 = 0, (\Delta_r y_n) \in l_p(\mathbb{Z}_+; X) \right\}, \ 1 < p < +\infty$ (Sect. 5.5).

$l_{p,r}^2(\mathbb{Z}_+; X) \equiv \left\{ (y_n) : y_0 = y_1 = 0, (\Delta_r^2 y_n) \in l_p(\mathbb{Z}_+; X) \right\}, \ 1 < p < +\infty$
(Sect. 5.5).

$l_{p,I-T}(\mathbb{Z}_+; X) \equiv \left\{ (y_n) : ((I - T) y_n) \in l_p(\mathbb{Z}_+; X) \right\}, \ 1 < p < +\infty.$

$l_1 \subset_c X \equiv$ This means that X contains l_1 as a complemented subspace.

$D(B) \equiv$ Denote the domain of a linear operator B on X. We consider $D(B)$ as a Banach space endowed with the graph norm.

$\rho(B) \equiv$ Denotes the resolvent of operator B and for $\lambda \in \rho(B)$, we write $R(\lambda, B) = (\lambda I - B)^{-1}$.

$\sigma(B) \equiv$ Denotes the spectrum of operator B.

$\sum_\delta \equiv$ Denotes the open sector $\{\lambda \in \mathbb{C} : |arg(\lambda)| < \delta\}$.

$co(\mathcal{T})$ (resp. $abco(\mathcal{T})$) $\equiv$ Denotes the convex hull (resp. absolute convex hull) of the set $\mathcal{T}$.

$\chi_A \equiv$ The characteristic function of the set A.

$$sign(t) \equiv \begin{cases} \dfrac{t}{|t|}, \ t \ne 0, \\ \\ 0, \ t = 0. \end{cases}$$

$r_j \equiv$ The jth Rademacher function, that is, $r_k(t) := sign(\sin(2^k \pi t))$. For $x \in X$ we denote by $r_j \otimes x$ the vector-valued function $t \to r_j(t) x$ (Sect. 3.2).

For $x \in X$ we denote by $e_k \otimes x$ the vector-valued function $t \to e_k(t) x$, where $t \in \mathbb{R}$, $k \in \mathbb{Z}$, and $e_k(t) = e^{ikt}$ (Sect. 2.4).

$\mathcal{F}f, \hat{f} \equiv$ The Fourier transform of f.

$\mathcal{F}^{-1}f, \check{f} \equiv$ The inverse Fourier transform of f.

$\mathcal{F}x(z), \hat{x}(z) \equiv$ Denotes the discrete-time Fourier transform of the sequence $x(n)$, that is, $\mathcal{F}x(z) = \hat{x}(z) = \sum_{n \in \mathbb{Z}} z^{-n} x(n)$.

$Z[x(n)], \tilde{x}(z) \equiv$ The Z-transform of a vector-valued sequence $x(n) \in X$ (or operator-valued sequence $x(n) \in \mathcal{B}(X)$), that is, $\tilde{x}(z) = Z[x(n)] = \sum_{j=0}^{\infty} z^{-j} x(j)$ (Sect. 1.2).

$a * y \equiv$ The convolution $*$ of a real, complex, or operator-valued sequence $a(n)$ and a vector-valued sequence $y(n)$.

$\mathcal{C}(n) \equiv$ The discrete-time cosine operator sequence (Sect. 1.4).

$\mathcal{S}(n) \equiv$ The discrete-time sine operator sequence (Sect. 1.4).

$M(n \times n) \equiv$ The set of all $n \times n$ matrices (Sect. 4.2).

$\mathcal{S}(\mathbb{R}^n; X) \equiv$ Denotes the Schwartz class of infinitely differentiable, rapidly decreasing X-valued functions.

$\mathcal{S}'(\mathbb{R}^n; X) \equiv$ Denotes the X-valued tempered distributions.

$H^p(\mathbb{R}^n; X) \equiv \{\mathcal{S}' - \sum_{k=0}^{\infty} \lambda_k a_k : a_k \ H^p - atom, \lambda_k \in \mathbb{C}, \sum_{k=0}^{\infty} |\lambda_k|^p < \infty\}$
(The atomic Hardy space). Equipped with the p-norm

$$\|f\|_{H^p(\mathbb{R}^n; X)}^p := \inf \sum_{k=0}^{\infty} |\lambda_k|^p,$$

where the infimum is taken over all atomic decompositions of $f \in H^p$ as above. The definition of the atoms appearing here is the same as in the scalar-valued context (see [101] for details).

$H^p([0, \infty); X) \equiv$ The subspace of $H^p(\mathbb{R}; X)$ consisting of all f whose distributional support lies on $[0, \infty)$ (see Sect. 2.5).

$\mathcal{D}([0, 2\pi]) \equiv$ The space of all complex-valued infinitely differentiable functions on $[0, 2\pi]$ equipped with the usual locally convex topology by the seminorms

$$\|f\|_k = \sup\{\|f^{(k)}(x)\| : 0 \le x \le 2\pi\},$$

for $k \in \mathbb{Z}_+$ (see Sect. 2.5).

$\mathcal{D}'([0, 2\pi]; X) \equiv$ The space of all linear continuous mappings from $\mathcal{D}([0, 2\pi])$ into X, that is, all X-valued distributions on $[0, 2\pi]$ (see Sect. 2.5).

$B_{p,q}^s([0, 2\pi]; X) \equiv$ Denotes the periodic Besov space (see Sect. 2.5).

$F_{p,q}^s([0, 2\pi]; X) \equiv$ Denotes the X-valued periodic Triebel–Lizorkin space (see Sect. 2.5).

$W_0^{\infty,p} \equiv$ The Banach space of all sequences $V = (V_n)$ belonging to $l_\infty(\mathbb{Z}_+, X)$ such that $V_0 = 0$ and $\Delta V \in l_p(\mathbb{Z}_+, X)$ equipped with the norm $\|V\|_{W_0^{\infty,p}} = \|V\|_\infty + \|\Delta V\|_p$ (Sect. 4.1).

$W_m^{\infty,p} \equiv$ The Banach space of all sequences $V = (V_n)$ belonging to $\ell_\infty(\mathbb{Z}_+, X)$ such that $V_n = 0$ if $0 \le n \le m$, and $\Delta V \in \ell_p(\mathbb{Z}_+, X)$ equipped with the norm $\|\cdot\|_{W_0^{\infty,p}}$ (Sect. 4.2).

$W_m^{\infty,p}[a] \equiv$ The closed ball $\|V\|_{W_0^{\infty,p}} \le a$ in $W_m^{\infty,p}$.

$MR_p(\mathbb{Z}_+; X) \equiv$ Denotes the space of maximal regularity (Sect. 5.6).

$C^\alpha(\mathbb{R}; X) \equiv$ Denotes the space of all X-valued functions u on $\mathbb{R}$, such that

$$\|u\|_\alpha := \sup_{s \ne t} \frac{\|u(s) - u(t)\|}{|s - t|^\alpha} < \infty.$$

We put $\|u\|_{C^\alpha} := \|u(0)\| + \|u\|_\alpha$. Then $C^\alpha(\mathbb{R}; X)$ is a Banach space under the norm $\|u\|_{C^\alpha}$ (Sect. 5.7).

$H^{1,p}([0, 2\pi]; X) \equiv$ The set of all $u \in L^p([0, 2\pi]; X)$ $(1 \le p < \infty)$ such that there is $v \in L^p([0, 2\pi]; X)$ satisfying $\int_0^{2\pi} v(t)dt = 0$ and, for some $x \in X$,

$$u(t) = x + \int_0^t v(t)ds$$

a.e. on $[0, 2\pi]$ (this is equivalent to saying that $\hat{v}(k) = ik\hat{u}(k)$ for all $k \in \mathbb{Z}$ (see [8])). Each element $u \in H^{1,p}([0, 2\pi]; X)$ has a continuous representative u such that $u(0) = u(2\pi)$, and u is a.e. differentiable on $[0, 2\pi]$ (see Sect. 5.7).

$H^{2,p}([0, 2\pi]; X) \equiv$ The set of all $u \in H^{1,p}([0, 2\pi]; X)$ such that $u' \in H^{1,p}([0, 2\pi]; X)$. Thus every $u \in H^{2,p}([0, 2\pi]; X)$ is twice differentiable a.e. on $[0, 2\pi]$, $u, u', u'' \in L^p([0, 2\pi]; X)$ and $u(0) = u(2\pi)$, $u'(0) = u'(2\pi)$ (see Sect. 5.7).

$\mathcal{H}_0^{2,p} \equiv$ The Banach space of all sequences $V = (V_n)$ such that $V_0 = V_1 = 0$ and $\Delta^2 V \in l_p(\mathbb{Z}_+; X)$ equipped with the norm $|||V||| = ||\Delta^2 V||_p$ (Sect. 6.2).

$\mathcal{H}_m^{2,p} \equiv$ The Banach space of all sequences $V = (V_n)$ such that $V_n = 0$ if $0 \leq n \leq m$, and $\Delta^2 V \in \ell_p(\mathbb{Z}_+; X)$ equipped with the norm $|||V||| = ||\Delta^2 V||_p$. For $\lambda > 0$, denote by $\mathcal{H}_m^{2,p}[\lambda]$ the ball $|||V||| \leq \lambda$ in $\mathcal{H}_m^{2,p}$ (Sect. 6.5).

$\mathcal{B}^{\omega,r} \equiv$ The Banach space of all X-valued sequences $y = (y_n)_{n \in \mathbb{Z}_+}$ such that

$$||y||_\omega := \sup_{n \in \mathbb{Z}_+} \left(\frac{||y_n||_X}{c(r,n)} + \frac{||\Delta_r y_n||_X}{c(r, n+1)} \right) < +\infty$$

(see Sect. 7.5).

$\mathcal{B}_m^{\omega,r} \equiv$ The Banach space of all sequences $y = (y_n)_{n \in \mathbb{Z}_+}$ belonging to $\mathcal{B}^{\omega,r}$ such that $y_n = 0$ if $0 \leq n \leq m$ equipped with the norm $|| \cdot ||_\omega$ (see Sect. 7.5).

$\mathcal{B}_\infty^{\omega,r} \equiv$ The Banach space of all weighted convergent functions $\xi \in \mathcal{B}_m^{\omega,r}$, that is, for which the limit

$$Z_\infty^{\omega,r}(\xi) := \lim_{n \to \infty} \frac{\xi(n)}{c(r,n)}$$

exists, endowed with the norm $|| \cdot ||_\omega$ (see (7.5.5) and Sect. 7.6).

$\mathcal{P}_{ps} \equiv$ Denotes a phase space defined axiomatically (see Sect. 7.7).

$l^p(\mathbb{Z}^+; \mathcal{P}_{ps}) \equiv \{\xi : \mathbb{Z}^+ \to \mathcal{P}_{ps} / ||\xi||_p^p := \sum_{n=0}^\infty ||\xi(n)||_{\mathcal{P}_{ps}}^p < \infty\}$, $1 \leq p < \infty$ (see Sect. 7.7).

$l^\infty(\mathbb{Z}^+; \mathcal{P}_{ps}) \equiv \{\xi : \mathbb{Z}^+ \to \mathcal{P}_{ps} / ||\xi||_\infty := \sup_{n \in \mathbb{Z}^+} ||\xi(n)||_{\mathcal{P}_{ps}} < \infty\}$ (see Sect. 7.7).

$\mathcal{P}_{ps}^\gamma \equiv \left\{\varphi : \mathbb{Z}^- \to \mathbb{C}^r : \sup_{\theta \in \mathbb{Z}^-} \frac{|\varphi(\theta)|}{e^{-\gamma\theta}} < \infty\right\}$ (see Sect. 7.7).

$\mathcal{P}_{ps}^{\beta,r} = \mathcal{P}_{ps}^{\beta,r}(\mathbb{Z}^-; \mathbb{C}^r) \equiv \{\varphi : \mathbb{Z}^- \longrightarrow \mathbb{C}^r : \mathrm{Sup}_{n \in \mathbb{Z}_+} |\varphi(-n)|/\beta(n) < +\infty\}$ (see Sect. 7.9).

$\mathcal{P}_{ps}^\circ(\tau) \equiv \left\{\varphi \in \mathcal{P}_{ps}^\gamma : \sum_{n=\tau}^\infty ||T(n, \tau)\varphi||_{\mathcal{P}_{ps}^\gamma}^p < +\infty\right\}$ (The l_p-stable space, see Sect. 7.10).

$$E^0(t) \equiv \begin{cases} I \ (r \times r \text{ unit matrix}), & t = 0, \\ 0 \ (r \times r \text{ zero matrix}), & t < 0. \end{cases}$$

$\Gamma(n, s) \equiv$ Green's function associated with (7.7.2), that is,

$$\Gamma(t, s) = \begin{cases} T(n, s + 1) P(s + 1) & n - 1 \geq s, \\ -T(n, s + 1) Q(s + 1) & s > n - 1. \end{cases}$$

$\mathcal{L}_m^p \equiv$ The closed subspace of $l^p(\mathbb{Z}^+, \mathcal{P}_{ps})$ of the sequences $\xi = (\xi(n))$ such that $\xi(n) = 0$ if $0 \leq n \leq m$. For $\lambda > 0$, denote by $\mathcal{L}_m^p[\lambda]$ the ball $\|\xi\|_p \leq \lambda$ in $\mathcal{L}_m^p$ (see Sect. 7.7).

$l_\varrho^\infty(\mathbb{Z}^+; \mathcal{P}_{ps}) \equiv \{\xi \colon \mathbb{Z}^+ \to \mathcal{P}_{ps} \ / \ \|\xi\|_\varrho = \sup_{n \in \mathbb{Z}^+} \|\xi(n)\|_{\mathcal{P}_{ps}} e^{-\varrho n} < \infty\}, \varrho > 0$ (see Sect. 7.8).

$l_\varrho^1(\mathbb{Z}^+; \mathbb{C}^r) \equiv \{\varphi \colon \mathbb{Z}^+ \to \mathbb{C}^r \ / \ \|\varphi\|_{1,\varrho} := \sum_{n=0}^\infty |\varphi(n)| e^{-\varrho n} < \infty\}, \varrho > 0$ (see Sect. 7.8).

Chapter 1
Discrete Semigroups and Cosine Operators

This chapter covers the basic elements of difference equations on Banach spaces and the tools necessary for studying and solving, primarily, vector-valued linear difference equations. It is motivated by a good number of examples from various fields, which are presented in the first section, and then discussed along with their detailed solutions in Chap. 7.

Difference equations are often used to model "an approximation" of differential equations, an approach which underlies the development of many numerical methods. However, there are many situations, for example, recurrence relations and the modeling of discrete processes such as traffic flow with finite number of entrances and exits, in which difference equations arise naturally. This justifies the use of discrete transforms on frequency domains. The discrete-time Fourier transform represents the main operational difference calculus method for solving difference equations associated with boundary conditions. On the other hand, the Z-transform is used to solve difference equations associated with initial values or "initial value problems". This is in parallel to how the Laplace transform is used for solving linear differential equations associated with initial conditions, i.e., the familiar initial value problems.

We present in Sect. 1.2 some elements of vector-valued Z-transform and discrete-time Fourier transform. Since a vector-valued theory follows literally the scalar-valued theory, we have only presented the necessary material needed for the use in the rest of the book. The main idea behind this section is to provide to the reader a powerful tool which helps to clarify the fundamental results that will appear in the subsequent sections. In this chapter we also present the basic discrete semigroup theory and introduce the discrete cosine and sine operators.

R.P. Agarwal et al., *Regularity of Difference Equations on Banach Spaces*, DOI 10.1007/978-3-319-06447-5_1, © Springer International Publishing Switzerland 2014

1.1 Difference Equations on Banach Spaces

In this section we consider several examples of difference equations arising in diverse contexts.[1]

The qualitative behavior of the solutions of second-order difference equations has been investigated by several authors; see e.g., [70, 72], the monographs [1, 75], and references therein.

The semilinear problem

$$
\begin{cases}
\Delta^2 x_n - (I - T)x_n = q_n f(x_n), n \in \mathbb{Z}_+, \\
\\
x_0 = 0, \; x_1 = 0,
\end{cases}
\tag{1.1.1}
$$

where f is defined on a Hilbert space H and $T \in \mathcal{B}(H)$ corresponds to a generalization of the equation

$$
\Delta^2 x_n = q_n f(x_n),
\tag{1.1.2}
$$

which was first considered by Drozdowicz and Popenda in [72]. There, they gave necessary and sufficient conditions for the existence of a solution which is asymptotically constant, that is, satisfies $\lim_{n \to \infty} x_n = c$, where c is a constant such that $f(c) \neq 0$. We note that an extension of this result for higher-order differential equations can be found in [71]. Furthermore, together with S. McKee, J. Popenda obtained conditions for the existence of asymptotically constant solutions for a system of linear difference equations [115].

The asymptotic behavior for the following scalar evolution problem

$$
\begin{cases}
x_{n+2} - \dfrac{3}{2}x_{n+1} + \dfrac{1}{2}x_n = \dfrac{e^{-n}}{1 + x_n^2}, \; n \in \mathbb{Z}_+, \\
\\
x_0 = x_1 = 0.
\end{cases}
\tag{1.1.3}
$$

is considered in Elaydi's book [75, Example 7.41, p. 347]. It corresponds to a second-order nonlinear equation of Poincare's type. Elaydi shows that the above equation has two solutions $x_1(n) \sim 1$ and $x_2(n) \sim (1/2)^n$. In Chap. 6, we will investigate for solutions belonging to vector-valued Lebesgue spaces.

Suppose now that $f = (f_n)$ belongs to some vector-valued Lebesgue space. Does the scalar difference equation

---

[1]We denote x_n or $x(n)$ for a sequence $x(\cdot)$ evaluated at the point n. However in Sects. 7.7–7.9, x_n represents the history function in the context of functional difference equations with infinite delay.

$$\begin{cases} x_{n+2} - x_{n+1} + \dfrac{1}{4}\sin(x_n)x_n = f_n, n \in \mathbb{Z}_+, \\ \\ x_0 = 0, \ x_1 = 0, \end{cases} \qquad (1.1.4)$$

have a solution in some space of maximal regularity?

The semilinear discrete control system

$$x_{n+1} = Ax_n + Bu_n + F(x_n, u_n), \quad n \in \mathbb{Z}_+, \qquad (1.1.5)$$

where A and B are constant matrices, F is a nonlinear function, and u_n a control input, was considered in [157] and consists of a linear discrete-time system and a bounded nonlinear perturbation. We ask for stable solutions of this system. In particular, in Chap. 7, we will be interested in the regularity of the semilinear difference equation

$$x_{n+1} = \alpha x_n + \beta u_n + c x_n \sin u_n + d u_n \cos^2 x_n, \quad n \in \mathbb{Z}_+. \qquad (1.1.6)$$

In the article [42], Cieśliński and Ratkiewicz discussed the discretizations of the second-order linear ordinary differential equations with constant coefficients. Special attention is given to the exact discretization because there is a difference equation whose solutions exactly coincide with solutions of the corresponding differential equation evaluated at a discrete sequence of points. Such exact discretization can be found for the harmonic oscillator equation $\ddot{x} + x = 0$ and is given by

$$x_{n+2} - 2\cos(\epsilon)x_{n+1} + x_n = 0. \qquad (1.1.7)$$

We consider in this monograph extensions of this discretization to Banach spaces and study existence and regularity of their solutions.

1.2   The Transform Method

In this book, we use the so-called frequency domain analysis or transform method to sometimes justify, and otherwise motivate, our techniques and results. In this approach, we transform the difference equations in the complex domain and then manipulate the resulting equations algebraically.

The transform method is most suitable for linear difference equations and discrete systems. It is widely used in the analysis and design of digital control, communications, and signal processing.

Let X be a Banach space. The Z-*transform* of a vector-valued sequence $x(n) \in X$ (or operator-valued sequence $x(n) \in \mathcal{B}(X)$) which is identically zero for negative integers n is defined by

$$\tilde{x}(z) \equiv Z[x(n)] = \sum_{j=0}^{\infty} z^{-j} x(j) \tag{1.2.1}$$

where z is a complex number.

The set of numbers z in the complex plane for which the series (1.2.1) converges is called the region of convergence of $x(z)$. The most commonly used method to find the region of convergence of the series (1.2.1) is the ratio test. Suppose that

$$\lim_{j \to \infty} \frac{||x(j+1)||}{||x(j)||} = R.$$

Then by the ratio test, the series (1.2.1) converges in the region $|z| > R$ and diverges for $|z| < R$. The number R is called the radius of convergence of series (1.2.1).

The Z-transform of some elementary functions is given in the following:

Example 1.2.1. Let T be a bounded operator defined on a Banach space X. Then

(i) $Z[T^n] = z(z - T)^{-1}$ for $|z| > \|T\|$.
(ii) $Z[nT^n] = zT(z - T)^{-2}$ for $|z| > \|T\|$.
(iii) $Z[n^2 T^n] = zT(z + T)(z - T)^{-3}$ for $|z| > \|T\|$.

Some useful properties of the Z-transform which will be needed in the sequel are stated in the following result.

Proposition 1.2.2. *The following properties hold:*

(a) *(Linearity) Let $\tilde{x}(z)$ be the Z-transform of $x(n)$ with radius of convergence R_1 and $\tilde{y}(z)$ be the Z-transform of $y(n)$ with radius of convergence R_2. Then for any complex numbers a, b, we have*

$$Z[ax(n) + by(n)] = a\tilde{x}(z) + b\tilde{y}(z) \quad for\ |z| > max\{R_1, R_2\}. \tag{1.2.2}$$

(b) *(Right shifting) Let R be the radius of convergence of $\tilde{x}(z)$. If $x(-i) = 0$ for $i = 1, 2, \ldots, k$, then*

$$Z[x(n - k)] = z^{-k} \tilde{x}(z), \ for\ |z| > R. \tag{1.2.3}$$

(c) *(Left shifting) Let R be the radius of convergence of $\tilde{x}(z)$. Then*

$$Z[x(n + k)] = z^k \tilde{x}(z) - \sum_{r=0}^{k-1} x(r) z^{k-r}, \ for\ |z| > R. \tag{1.2.4}$$

In particular

$$Z[x(n + 1)] = z\tilde{x}(z) - zx(0), \ for\ |z| > R, \tag{1.2.5}$$

and

$$Z[x(n+2)] = z^2 \tilde{x}(z) - z^2 x(0) - zx(1), \; for \; |z| > R. \qquad (1.2.6)$$

(d) *(Initial value theorem)*

$$\lim_{|z|\to\infty} \tilde{x}(z) = x(0). \qquad (1.2.7)$$

(e) *(Final value theorem)*

$$x(\infty) = \lim_{n\to\infty} x(n) = \lim_{z\to 1}(z-1)\tilde{x}(z). \qquad (1.2.8)$$

(f) *(Convolution) A convolution $*$ of a real or complex-valued sequence $a(n)$ and a vector-valued sequence $y(n)$ is defined by*

$$(a * y)(n) \equiv a(n) * y(n) = \sum_{j=0}^{n} a(n-j)y(j). \qquad (1.2.9)$$

Now

$$Z[a(n) * y(n)] = \tilde{a}(z)\tilde{y}(z). \qquad (1.2.10)$$

The same formula holds if the convolution is defined as

$$a(n) * y(n) = \sum_{j=0}^{\infty} a(n-j)y(j). \qquad (1.2.11)$$

(g) *(Multiplication by a^n) Suppose that $\tilde{x}(z)$ is the Z-transform of $x(n)$ with radius of convergence R. Then*

$$Z[a^n x(n)] = \tilde{x}(z/a), \quad for \; |z| > |a|R. \qquad (1.2.12)$$

(h) *(Multiplication by n^k)*

$$Z[n^k x(n)] = (z\frac{d}{dz})^k Z[x(n)]. \qquad (1.2.13)$$

(i) *(Uniqueness) Suppose that there are two vector-valued sequences $x(n)$, $y(n)$ such that $\tilde{x}(z) = \tilde{y}(z)$ for $|z| > R$. Then $x(n) \equiv y(n)$.*

Example 1.2.3. Let T be a bounded operator defined on a Banach space X. Then $Z[T^{n+1}] = z^2(z-T)^{-1} - zI$ for $|z| > \|T\|$.

Consider a circle C, centered at the origin of the complex plane, that encloses all poles of $\tilde{x}(z)z^{n-1}$. Then by the Cauchy's integral formula the following expression for the inverse Z-transform follows:

$$Z^{-1}[\tilde{x}(z)] \equiv x(n) = \frac{1}{2\pi i} \int_C \tilde{x}(z)z^{n-1}dz$$

$$= \text{sum of residues of } \tilde{x}(z)z^{n-1}.$$

Given a vector-valued sequence $x(n)$, the *discrete-time Fourier transform* on $\mathbb{T} := \{z \in \mathbb{C} : |z| = 1\}$ is defined as

$$\mathcal{F}x(z) = \hat{x}(z) = \sum_{n \in \mathbb{Z}} z^{-n}x(n), \quad z \in \mathbb{T}.$$

Alternatively, we write

$$\mathcal{F}x(\omega) = \hat{x}(\omega) = \sum_{n \in \mathbb{Z}} x(n)e^{-in\omega}.$$

The notation $\hat{x}(z)$ helps to highlight the periodicity property and emphasizes the relationship of the discrete-time Fourier transform to the Z-transform.

Vector-valued discrete-time Fourier transform arises naturally when one considers communication problems across a multiple-input multiple-output channel (MIMO communications) (see [16]).

The discrete-time Fourier transform is Pontryagin dual to the Fourier series, which transforms from a periodic domain to a discrete domain. It provides an approximation of the continuous-time Fourier transform. The discrete-time Fourier transform plays a key role in representing and analyzing discrete-time signals and systems.

We note that the convolution theorem for the discrete-time Fourier transform holds, i.e., $\widehat{x * y}(z) = \hat{x}(z)\hat{y}(z)$. Here, the convolution is defined in analogous way to the case of Z-transform:

$$(x * y)(n) = \sum_{m=-\infty}^{\infty} x(m)y(n - m).$$

Further properties of the discrete-time Fourier transform are analogous to those of the Z-transform, since it is the evaluation of the Z-transform around the unit circle in the complex plane.

The following inverse transforms recover the discrete-time sequence:

$$x(n) = \frac{1}{2\pi} \int_{-\pi}^{\pi} \hat{x}(\omega)e^{in\omega}d\omega.$$

1.3 Discrete Semigroups Operators

Discrete-time semigroups are sequences $\{\mathcal{T}(n)\}_{n\in\mathbb{Z}_+} \subset \mathcal{B}(X)$ satisfying

$$\mathcal{T}(n+m) = \mathcal{T}(n)\mathcal{T}(m), \quad n, m \in \mathbb{Z}_+,$$

and

$$\mathcal{T}(0) = I.$$

Each discrete-time semigroup is uniquely determined by the single value of $\mathcal{T}(1)$. Indeed, we have $\mathcal{T}(n+1) = \mathcal{T}(n)\mathcal{T}(1)$ for all $n \in \mathbb{Z}_+$, and hence it follows by induction that $\mathcal{T}(n) = \mathcal{T}(1)^n$. We define

$$\Delta\mathcal{T}(0) := \mathcal{T}(1) - \mathcal{T}(0).$$

This element will be called the *generator* of $\{\mathcal{T}(n)\}_{n\in\mathbb{Z}_+}$, by analogy to the continuous case.

For $T \in \mathcal{B}(X)$ given, define $\mathcal{T} : \mathbb{Z}_+ \to \mathcal{B}(X)$ by

$$\mathcal{T}(n) = T^n.$$

Then $\mathcal{T}$ is a discrete-time semigroup with generator $T - I$. Note that there exists a bijection between the set $\mathcal{B}(X)$ and the set of all discrete-time semigroups.

In what follows, our concern is in the discrete-time evolution equation of first order:

$$\begin{cases} \Delta x_n - (T - I)x_n = f_n, \ n \in \mathbb{Z}_+, \\ \\ x_0 = x, \end{cases} \tag{1.3.1}$$

or, equivalently,

$$\begin{cases} x_{n+1} - T x_n = f_n, \ n \in \mathbb{Z}_+, \\ \\ x_0 = x, \end{cases} \tag{1.3.2}$$

where the sequence $f = (f_n)$ is given.

Let $\{S(n)\}_{n\in\mathbb{Z}_+} \subset \mathcal{B}(X)$ and $f : \mathbb{Z}_+ \to X$ be a vector-valued sequence. In what follows, we denote [see (1.2.9)]

$$(S * f)_n := \sum_{j=0}^{n} S(n - j)f(j).$$

Proposition 1.3.1. *Let $T \in \mathcal{B}(X)$. The unique solution of equation (1.3.1) is given by*

$$x_{n+1} = T(n+1)x + (\mathcal{T} * f)_n. \tag{1.3.3}$$

Moreover

$$\Delta x_{n+1} = (T-I)T(n)x + (T-I)(\mathcal{T} * f)_n + f_{n+1}. \tag{1.3.4}$$

Proof. Note that the Z-transform of (1.3.2) gives

$$(z - T)\tilde{x}(z) = z x_0 + \tilde{f}(z),$$

equivalently

$$\tilde{x}(z) = z(z - T)^{-1}x_0 + (z - T)^{-1}\tilde{f}(z),$$

whenever $z \in \rho(T)$. This is equivalent to

$$z\tilde{x}(z) - z x_0 = z(z - T)^{-1}x_0 - z x_0 + z(z - T)^{-1}\tilde{f}(z).$$

Hence, observing the properties of the Z-transform [see Example 1.2.3(i)], we conclude that the (unique) solution of (1.3.2) is given by

$$x_{n+1} = T(n+1)x + (\mathcal{T} * f)_n. \tag{1.3.5}$$

In particular, in the case of $x = 0$, we obtain from (1.3.5)

$$\Delta x_{n+1} = \sum_{k=0}^{n}(T-I)T^{n-k}f_k + f_{n+1} = \sum_{k=0}^{n}(T-I)T^k f_{n-k} + f_{n+1}. \tag{1.3.6}$$

$\square$

Denote $A := T - I$. The continuous counterpart of equation (1.3.1) is the evolution equation

$$\begin{cases} u'(t) - Au(t) = f(t), & t \geq 0, \\[2mm] u(0) = x, \end{cases} \tag{1.3.7}$$

where $f : \mathbb{R}^+ \to X$ is given. Its solution is provided by the variation of parameters formula:

$$u(t) = e^{At}x + \int_0^t e^{A(t-s)}f(s)ds, \quad t \geq 0. \tag{1.3.8}$$

Comparing with the discrete case for $f \equiv 0$ it becomes clear that the continuous counterpart of the discrete semigroup is the exponential function:

$$S(t) := e^{At} = e^{(T-I)t}.$$

A well-known problem for continuous-time evolution equations is the following: We consider the evolution equation (1.3.7) with initial condition $u(0) = 0$ and one looks for the solution

$$u(t) = \int_0^t e^{A(t-s)} f(s)ds \equiv (S(\cdot) * f)(t),$$

on $\mathbb{R}^+$. Then one says that (1.3.7) has *maximal regularity* if the map

$$f \to \frac{dS}{dt} * f$$

defines a bounded operator on $L^p(\mathbb{R}^+; X)$ for all $p \in (1, \infty)$.

In 2001, Weis [180] proved a characterization of maximal regularity, which in Hilbert spaces reads as follows.

Theorem 1.3.2. *Let H be a Hilbert space and let e^{tA} be bounded and analytic on H (see Sect. 1.5). Then the following are equivalent:*

(i) *Equation (1.3.7) has maximal regularity.*
(ii) *$\{z(z - A)^{-1} : z \in i\mathbb{R}, z \neq 0\}$ is bounded.*
(iii) *$\{e^{tA}, tAe^{tA} : t \geq 0\}$ is bounded.*

One of the main objectives in Chap. 3 will be to show the discrete counterpart of the above result. The following concepts, which correspond to the discrete version of part (iii) in Weis' theorem, will be frequently used in this book.

Definition 1.3.3. An operator $T \in \mathcal{B}(X)$ is said to be power bounded if the set

$$\{\|T^n\| : n \in \mathbb{N}\}$$

is bounded.

In other words, an operator T is bounded if and only if $\mathcal{T} \in l_\infty(\mathbb{Z}_+; \mathcal{B}(X))$.

Remark 1.3.4. If $T \in \mathcal{B}(X)$ is similar to a contraction, that is, if there is an invertible $U \in \mathcal{B}(X)$ with $\|UTU^{-1}\| \leq 1$, then T is power bounded. The converse is not true; a power-bounded operator in $\mathcal{B}(X)$ need not be similar to a contraction (see [80]).

Remark 1.3.5. $T \in \mathcal{B}(X)$ is power bounded if and only if

$$\limsup_{n \to \infty} \|T^n x\| < \infty,$$

for all $x \in X$.

Remark 1.3.6. Let T be power bounded on X; then

$$||R(\lambda, I - T)|| \leq \frac{M(T)}{|\lambda - 1| - 1} \leq \frac{M(T)}{|Re(\lambda)|}, \quad Re(\lambda) > 0,$$

where $M(T) := \sup_{n \in \mathbb{Z}^+} ||T^n||$ (see [94]).

The following definition was introduced by Coulhon and Saloff-Coste in [49].

Definition 1.3.7. An operator $T \in \mathcal{B}(X)$ is called analytic if the set

$$\{n(T - I)T^n : n \in \mathbb{N}\}$$

is bounded.

Remark 1.3.8. The above condition does not necessarily imply that T is power bounded. In fact, there is a bounded operator T on $L^1(\mathbb{R})$ such that $\sup_n n||T^{n+1} - T^n|| < \infty$, and $||T^n|| \approx \log n$ (see [110]).

Theorem 1.3.9. *Let $T \in \mathcal{B}(X)$. The following assertions are equivalent:*

(a) *T is power bounded and analytic.*
(b) *$e^{t(T-I)}$ is bounded and analytic and $\sigma(T) \subset \{z \in \mathbb{C} : |z| < 1\} \cup \{1\}$.*

The question of the relationship between boundedness and analyticity of discrete semigroups has been deeply studied. In this direction we give the following remark:

Remark 1.3.10. Let X be Banach space and $T \in \mathcal{B}(X)$.

(i) If T is power bounded and $\sigma(T) \cap \mathbb{T} = \{1\}$, then

$$\lim_{n \to \infty} ||T^{n+1} - T^n|| = 0$$

   (see [78, 114]).
(ii) If $L := \limsup_{n \to \infty} ||n(T^{n+1} - T^n)|| < 1/e$, then T is power bounded, but it does not necessary hold if $L = 1/e$. If $\sigma(T) = \{1\}$ and $T \neq I$, then $\liminf_{n \to \infty} ||n(T^{n+1} - T^n)|| \geq 1/e$. The constant $1/e$ is sharp (see [110]).
(iii) T is power bounded and analytic if and only if $\{(\lambda - 1)R(\lambda, T) : \lambda \in \overline{\mathbb{D}}^c \cup (1 + \sum_\delta)\}$ is bounded for some $\delta > \pi/2$. Here $\mathbb{D}$ is the unit disk in $\mathbb{C}$ and $\sum_\delta$ denotes the open sector $\{z : |arg(z)| < \delta\}$ (see [22]).

Blunck in [22, 23] gives many applications of analyticity condition to maximal regularity. For recent and related results on analytic operators we refer the reader to Dungey [73].

1.4   Discrete Cosine and Sine Operators

Let $T \in \mathcal{B}(X)$. A *discrete-time cosine operator sequence* is the unique solution of the second-order difference equation

$$\begin{cases} \Delta^2 u(n) = (I - T)u(n), \ n \in \mathbb{Z}_+, \\ \quad u(0) = x, \\ \quad u(1) = x. \end{cases} \tag{1.4.1}$$

We denote this unique solution by $\mathcal{C}(n)x$ $(n \in \mathbb{Z}_+)$, $x \in X$. We define $\mathcal{C}(n) = 0$ for $n = -1, -2, \ldots$. It follows from the definition that $\mathcal{C}(n) \in \mathcal{B}(X)$. The *generator* of $\mathcal{C}$ is the operator $I - T$.

Remark 1.4.1. Note that the definition of second-order difference operator is not unique. In consequence different notions of cosine operator sequences appear. In this section, we consider $\Delta^2 u(n) := \Delta(\Delta u)(n)$. However, as we will see in the next chapters, there are other possibilities more related to numerical aspects.

We define the *discrete-time sine operator sequence* as the unique solution of the difference equation

$$\begin{cases} \Delta^2 u(n) = (I - T)u(n), \ n \in \mathbb{Z}_+, \\ \quad u(0) = 0, \\ \quad u(1) = x. \end{cases} \tag{1.4.2}$$

We denote this unique solution by $\mathcal{S}(n)x$ $(n \in \mathbb{Z}_+)$. We also define $\mathcal{S}(n) = 0$ for negative n.

The following proposition gives explicit formulas for the discrete-time Fourier transform of the discrete sine and cosine operators. Note that by the definitions of $\mathcal{C}$ and $\mathcal{S}$ for negative n, their discrete-time Fourier transform coincides with the Z-transform on $\mathbb{T} := \{z \in \mathbb{C} : |z| = 1\}$.

Proposition 1.4.2. *Let $T \in \mathcal{B}(X)$ and suppose $\{(z - 1)^2\}_{z \in \mathbb{T}} \subseteq \rho(I - T)$. Then*

$$\hat{\mathcal{S}}(z) = z\big((z - 1)^2 - (I - T)\big)^{-1}, \quad z \in \mathbb{T}, \tag{1.4.3}$$

and

$$\hat{\mathcal{C}}(z) = z(z - 1)\big((z - 1)^2 - (I - T)\big)^{-1}, \quad z \in \mathbb{T}. \tag{1.4.4}$$

Proof. We note that the Z-transform of the second-order difference equation

$$\Delta^2 x(n) = (I - T)x(n), n \in \mathbb{Z}_+,$$

gives the identity

$$\left(z^2 - 2z + T\right)\tilde{x}(z) = z(z - 2)x(0) + zx(1)$$

or, equivalently,

$$\tilde{x}(z) = z(z - 2)\left((z - 1)^2 - (I - T)\right)^{-1}x(0)$$

$$+ z\left((z - 1)^2 - (I - T)\right)^{-1}x(1)$$

$$= z(z - 1)\left((z - 1)^2 - (I - T)\right)^{-1}x$$

in case $x(0) = x(1) = x$, and

$$\tilde{x}(z) = z\left((z - 1)^2 - (I - T)\right)^{-1}x,$$

in case $x(0) = 0$ and $x(1) = x$. Inverting the transform in both cases, we obtain the formulas in Proposition 1.4.2.   □

Remark 1.4.3. The main advantage to define the operators C and S as solutions of difference equations is that in this way additional hypothesis on T is not necessary. For example, let $T \in \mathcal{B}(X)$ be given and suppose that $(I - T)^{1/2}$ exists then, using the Z-transform, it is not difficult to obtain the representations

$$C(n) = \frac{(I + (I - T)^{1/2})^n + (I - (I - T)^{1/2})^n}{2} \tag{1.4.5}$$

and

$$(I - T)^{1/2}S(n) = \frac{(I + (I - T)^{1/2})^n - (I - (I - T)^{1/2})^n}{2}. \tag{1.4.6}$$

Indeed, one should observe that the Z-transform of (1.4.5) is given by

$$\hat{C}(z) = z\left((z - 1) - (I - T)^{1/2}\right)^{-1} + z\left((z - 1) + (I - T)^{1/2}\right)^{-1}$$

$$= z(z - 1)\left((z - 1)^2 - (I - T)\right)^{-1}.$$

And analogously using the Z-transform for (1.4.6) we obtain

$$(I - T)^{1/2}\hat{S}(z) = z\left((z-1) - (I - T)^{1/2}\right)^{-1}$$

$$-z\left((z-1) + (I - T)^{1/2}\right)^{-1}$$

$$= z(I - T)^{1/2}\left((z-1)^2 - (I - T)\right)^{-1}.$$

However, we note that the existence of $(I - T)^{1/2}$ is necessary.

In what follows, we are interested in a representation of the solution for the second-order difference equation

$$
\begin{cases}
\Delta^2 x_n - (I - T)x_n = f_n, \ n \in \mathbb{Z}_+, \\
x_0 = x, \ \Delta x_0 = x_1 - x_0 = y,
\end{cases}
\tag{1.4.7}
$$

where $f : \mathbb{Z}_+ \to X$.

For our next result, we have to use properties of discrete-time cosine and sine operators with respect to the Δ operator. They are summarized in the following proposition.

Proposition 1.4.4. *We have the following properties:*

(i) $\Delta \mathcal{S}(n) = \mathcal{C}(n), \ n \in \mathbb{Z}_+,$
(ii) $\Delta \mathcal{C}(n) = (I - T)\mathcal{S}(n), \ n \in \mathbb{Z}_+,$
(iii) $(\mathcal{S} * \Delta f)_n = (\Delta \mathcal{S} * f)_n - \mathcal{S}(n + 1)f_0, \ f = (f_n)_{n \in \mathbb{Z}_+}.$ *In particular, if* $f_0 = 0$, *then*

$$\mathcal{S} * \Delta f = \Delta \mathcal{S} * f.$$

Proof.

(i) Multiplying the first identity in Proposition 1.4.2 by $(z - 1)$ we obtain the identity

$$(z - 1)\hat{S}(z) = \hat{C}(z)$$

or, equivalently,

$$\left(z\hat{S}(z) - z\mathcal{S}(0)\right) - \hat{S}(z) = \hat{C}(z)$$

showing (i) in the frequency domain.

(ii) Follows from the definition of discrete-time operator $\mathcal{S}$ since it satisfies $\Delta^2\mathcal{S}(n) = (I - T)\mathcal{S}(n)$ and (i).

(iii) Follows from the identity

$$\hat{S}(z)\Big(z\hat{f}(z) - zf(0)\Big) = \Big(z\hat{S}(z) - zS(0)\Big)\hat{f}(z) - \Big(z\hat{S}(z) - zS(0)\Big)f(0).$$

□

The main result in this section is the following representation of the solution for the second-order difference equation.

Proposition 1.4.5. *Let* $T \in \mathcal{B}(X)$ *be given, then the (unique) solution of equation (1.4.7) is given by*

$$x_{m+1} = C(m+1)x + S(m+1)y + (S * f)_m. \tag{1.4.8}$$

Moreover,

$$\Delta x_{m+1} = (I - T)S(m+1)x + C(m+1)y + (C * f)_m. \tag{1.4.9}$$

Proof. We give a simplified proof using the Z-transform. Indeed, taking the Z-transform to (1.4.7), we obtain the identity

$$\Big(z^2 - 2z + T\Big)\tilde{x}(z) = z(z-2)x(0) + zx(1) + \tilde{f}(z)$$

or, equivalently,

$$\tilde{x}(z) = z(z-2)\Big((z-1)^2 - (I - T)\Big)^{-1} x(0)$$

$$+ z\Big((z-1)^2 - (I - T)\Big)^{-1} x(1)$$

$$+ \Big((z-1)^2 - (I - T)\Big)^{-1} \tilde{f}(z),$$

whenever $(z-1)^2 \in \rho(I - T)$. Using the fact that $x(1) = y + x(0)$, we notice the equivalent formulation

$$z\tilde{x}(z) - zx_0 = z^2(z-1)\Big((z-1)^2 - (I - T)\Big)^{-1} x_0 - zx_0$$

$$+ z^2\Big((z-1)^2 - (I - T)\Big)^{-1} y \tag{1.4.10}$$

$$+ z\Big((z-1)^2 - (I - T)\Big)^{-1} \tilde{f}(z).$$

In order to know the value of $x(n + 1)$, which corresponds to the left-hand side of the above identity, we have to invert term by term the right-hand side.

Comparing the formulas of Proposition 1.4.2 with the right-hand side in the identity (1.4.10), we arrive easily to conclude the first part of the claim. For the second part, we apply Proposition 1.4.4. □

As a simple application, we are able to obtain the following functional equation for the sine function.

Corollary 1.4.6. $\mathcal{S}(m + n) = \mathcal{S}(n)\mathcal{C}(m) + \mathcal{S}(m)\mathcal{C}(n), \; n, m \in \mathbb{Z}_+.$

Proof. Fix $m \in \mathbb{Z}_+$, and $v \in X$. By the first part of Proposition 1.4.5, the unique solution of equation (1.4.7) with $f \equiv 0$ and initial conditions $x = \mathcal{S}(m)v$ and $y = \mathcal{C}(m)v$ is given by

$$z(n) = \mathcal{C}(n)x + \mathcal{S}(n)y = \mathcal{C}(n)\mathcal{S}(m)v + \mathcal{S}(n)\mathcal{C}(m)v. \quad (1.4.11)$$

But note that $w(n) := \mathcal{S}(n + m)v$ is also a solution of the same equation. It shows the assertion. □

1.5 Comments

Most of the material on the Z-transform of Sect. 1.2 follows the book of Elaydi [75]. We refer to this text for examples and proofs in the scalar case, which are identical in the vector-valued case. Three other very interesting books that include extensive treatment of the Z-transform method are [35, 108, 154].

Next, for further use, we recall the important notion of analytic semigroup [77].

Definition 1.5.1. Let $\delta \in (0, \frac{\pi}{2}]$ and $\sum_\delta = \{\lambda \in \mathbb{C} : |arg(\lambda)| < \delta\}$. A family of operator $(T_z)_{z \in \sum_\delta \cup \{0\}} \subset \mathcal{B}(X)$ is called an analytic semigroup (of angle δ) if

(i) $T(0) = I$ and $T(z + z') = T(z)T(z')$ for all $z, z' \in \sum_\delta \cup \{0\}$.
(ii) The map $z \to T(z)$ is analytic in $\sum_\delta \cup \{0\}$.
(iii) $\displaystyle\lim_{z \in \sum_{\delta'} \cup \{0\}, z \to 0} T(z)x = x$ for all $x \in X$ and $0 < \delta' < \delta$.

If in addition

(iv) $\{\|T(z)\| : z \in \sum_{\delta'} \cup \{0\}\}$ is bounded for each $0 < \delta' < \delta$,

we call $(T_z)_{z \in \sum_\delta \cup \{0\}}$ a bounded analytic semigroup.

We have the following characterization [77, Theorem 4.6]:

Theorem 1.5.2. *For a closed linear operator $(A, D(A))$ acting on a Banach space X the following assertions are equivalent.*

(a) *A generates a bounded analytic semigroup $(T_z)_{z \in \sum_\delta \cup \{0\}}$.*

(b) *There is $\theta \in (0, \frac{\pi}{2})$ such that the operators $e^{\pm i\theta} A$ generate bounded strongly continuous semigroups.*
(c) *A generates a bounded strongly continuous semigroup $(T_t)_{t \in \mathbb{R}^+}$ such that $Ran(T(t)) \subset D(A)$ for all $t > 0$ and $\sup_{t>0} \|tAT(t)\| < \infty$.*
(d) *A generates a bounded strongly continuous semigroup $(T_t)_{t \in \mathbb{R}^+}$ and there is a constant $C > 0$ such that $\|sR(r + is, A)\| \leq C$ for all $r > 0$ and all $s \in \mathbb{R}$.*
(e) *A is sectorial.*

Remark 1.5.3.

  (i) If A is a normal operator on a Hilbert space satisfying $\sigma(A) \subset \{z \in \mathbb{C} : arg(-z) < \delta\}$ for some $\delta \in [0, \frac{\pi}{2})$, then A generates a bounded analytic semigroup.
 (ii) Let A be the generator of a strongly continuous group; then A^2 generates an analytic semigroup of angle $\pi/2$.

The discrete formulation of the maximal regularity problem for the first-order equations was developed for Blunck in an illuminating paper [22] and indicated to him by T. Coulhon. The material on discrete semigroups is extracted from this paper (see Chap. 3).

The property of discrete analyticity is equivalent, in the case where T is power bounded, to the operator theoretical Ritt's condition:

$$\exists C > 0 \quad \|(1 - \lambda)R(\lambda, T)\| \leq C \quad \forall |\lambda| > 1. \tag{1.5.1}$$

The latter was proved independently by Nagy and Zemanek (see [152]) and Lyubich (see [138]) in 1999 and may be considered as an analogue of the equivalence between the notion of sectoriality for closed operators and the notion of analyticity for continuous-time semigroups (see, for instance, [77] and [164]).

The proof of Theorem 1.3.9 is given in [23, Theorem 2.3].

The following result was established by Blunck [23].

Theorem 1.5.4 ([23, Theorem 1.1]). *Let $p, q \in [1, \infty]$ and $T \in \mathcal{B}(L_p)$ be power bounded and analytic. If T is power bounded and analytic on L_q, then T is power bounded and analytic on L_r for all r strictly between p and q.*

The material on discrete sine and cosine functions is extracted from [53], where the authors began to develop the maximal regularity theory for the second-order discrete equations. The idea is to pursue the analogy with the continuous second-order Cauchy problem:

$$u''(t) = Au(t) + f(t).$$

However, using the delta operator, the resulting scheme is different from the discrete equation

$$x_{n+2} = Ax_n + f_n$$

which is more natural from the point of view of Z-transform, in the sense that we have only to replace $1/z$ by $1/z^2$ in the frequency domain. The advantage will be perceived when we study the case of exact discretization for second-order difference equations.

A rigorous proof of Proposition 1.4.2 is given in Cuevas and Lizama [53]. In this reference, sine and cosine operator functions are also explicitly given.

Not much is known for discrete cosine functions. The case of discrete semigroup is clear because of the existing bijection between the set of all bounded operators and the set of all discrete semigroups defined by means of Cauchy's functional equation. The correspondence with the set of all discrete cosine functions defined by means of D'Alembert functional equation remains an open problem.

It should be interesting to associate a functional equation to discrete cosine function defined by (1.4.1) when Δ^2 is defined, for example, by the relation $\Delta^2 u(n) = u(n+1) - u(n-1)$, $n \in \mathbb{Z}$.

Chapter 2
Maximal Regularity and the Method of Fourier Multipliers

Difference equations in a Banach space X of the form

$$\Delta u(n) = Au(n) + f(n) \tag{2.0.1}$$

arise in several branches of mathematical physics and engineering. In general, difference equations appear as natural descriptions of observed evolution phenomena because most measurements of time-evolving variables are discrete. Hence the applications of the theory of difference equations is rapidly increasing to several fields such as numerical analysis, control theory, finite mathematics, and computer science [103,130]. In particular there is a well-established basic theory for (2.0.1) in the scalar or matrix-valued case, which has been presented coherently, for example, in Agarwal's book [1] and Elaydi's book [75].

As we have mentioned in the preface, one of the objectives in this monograph is to show that maximal regularity is an important tool in the investigation of the existence and uniqueness of solutions to semilinear difference equations. Maximal regularity will be used in the forthcoming chapters to reduce the nonlinear problem, via a fixed-point argument,[1] to a linear problem.

One way to treat maximal regularity is to apply the discrete-time Fourier transform to the equation (2.0.1). The problem is now to decide whether the *operator-valued* function

$$M(z) = (z-1)\Big((z-1) - A\Big)^{-1}, |z| = 1, z \neq 1$$

---

[1]The usefulness of the fixed-point methods for applications has increased enormously by the development of efficient techniques for computing fixed points. In fact, nowadays, fixed-point arguments have become a powerful weapon in the arsenal of applied mathematicians [102].

R.P. Agarwal et al., *Regularity of Difference Equations on Banach Spaces*,
DOI 10.1007/978-3-319-06447-5_2, © Springer International Publishing Switzerland 2014

is a Fourier multiplier. This has been studied by Blunck in [22] who gave an equivalent property to maximal regularity of (2.0.1) in terms of bounds of the resolvent operator of A. To describe their result, we need to define the notions of UMD space and R-boundedness.

2.1 UMD Spaces

It is a general theme in functional analysis during the last decades, and in Banach space theory in particular, to investigate if classical results about scalar-valued functions remain valid if the functions considered take values in some Banach space. Usually, one of three things happens. The most desirable case having in mind applications is, if the results simply remain true, in the vector-valued setting for any Banach space. The worst case occurs if only trivial extensions remain true, possibly for functions with values in finite dimensional spaces or Hilbert spaces. The third and frequently observed case is that it depends on the structure and geometry of the Banach spaces considered and whether a result can be carried over to the vector-valued setting. The latter case often leads to a fruitful interplay between the geometry of Banach spaces and techniques from the classical theory. Prominent examples are the theory of type and cotype of Banach spaces initiated by Maurey and Pisier [143] (see Sect. 3.4) and the theory of UMD spaces developed by Burkhölder [30–32, 34] and Bourgain [24].

Definition 2.1.1. A Banach space X is said to have the unconditional martingale difference (UMD) property if for each $p \in (1, \infty)$ there is a constant $C_p > 0$ such that for any martingale $(f_n)_{n \geq 0} \subset L^p(\Omega, \Sigma, \mu; X)$ and any choice of signs $(\xi_n)_{n \geq 0} \subset \{-1, 1\}$ and any $N \in \mathbb{Z}_+$ the following estimate holds:

$$\left\| f_0 + \sum_{n=1}^{N} \xi_n (f_n - f_{n-1}) \right\|_{L^p(\Omega, \Sigma, \mu; X)} \leq C_p \| f_N \|_{L^p(\Omega, \Sigma, \mu; X)}.$$

Remark 2.1.2.

(i) A Banach space X is said to be $\mathcal{HT}$, if the Hilbert transform is bounded on $L^p(\mathbb{R}, X)$ for some (and then all) $p \in (1, \infty)$. Here, the Hilbert transform H of a function $f \in \mathcal{S}(\mathbb{R}, X)$, the Schwartz space of rapidly decreasing X-valued functions, is defined by

$$Hf(t) := PV - \int \frac{1}{t-s} f(s) ds.$$

It is well known that the set of Banach spaces of class $\mathcal{HT}$ coincides with the class of UMD spaces. This has been shown by Bourgain [24] and Burkhölder [32]. For more information and details on the Hilbert transform and the UMD Banach spaces, we refer to Amann's book [7, Sects. III.4.3–III.4.5].

 (ii) The UMD spaces include Hilbert spaces, Sobolev spaces $W_p^s(\Omega)$, $1 < p < \infty$
      (see [8]), Lebesgue spaces $L^p(\Omega, \mu)$, $L^p(\Omega, \mu; X)$, l_p, $1 < p < \infty$, where
      X is a UMD space, the reflexive noncommutative L^p spaces, Hardy spaces,
      Lorentz and Orlicz spaces, any von Neumann algebra, and the Schatten–von
      Neumann classes $C_p(H)$, $1 < p < \infty$, of operators on Hilbert spaces. On
      the other hand, the space of continuous functions $C(K)$ do not have the UMD
      property.
(iii) If $1 < p < \infty$ and the Lebesgue–Bochner space $L^p((0, 1), X)$ has an
      unconditional basis, then X is UMD [31].

Proposition 2.1.3 ([65]). *The following properties hold:*

  (i) *If X is UMD and Y is a closed linear subspace of X, then Y and X/Y are
      UMD spaces.*
 (ii) *A Banach space X is UMD if and only if its dual X^* is UMD.*
(iii) *A UMD space is always uniformly convex, but not conversely.*
 (iv) *If X and Y are UMD spaces, then $X \oplus Y$ is a UMD space.*
  (v) *Every UMD space is super-reflexive[2] but not conversely; see [24]. In particu-
      lar, L^1 and L^∞ are not UMD spaces.*

2.2 R-Boundedness

We note that the most important tools to prove maximal regularity are Fourier
multiplier theorems. They play a key role in the analysis of elliptic and parabolic
problems. In particular, in [17], the theory of operator-valued Fourier multiplier is
applied to obtain results on the hyperbolicity of delay equations. Also in [131],
it is applied to obtain stability of linear control systems in Banach spaces, and in
[117, 118], it is used to study maximal regularity for integrodifferential equations.

Classical theorems on L^p-multipliers are no longer valid for operator-valued
functions unless the underlying space is isomorphic to a Hilbert space. However,
work of Clément et al. [45], Weis [179, 180], and Clément-Prüss [48] show that
the right notion in this context is R-boundedness of sets of operators. Hence the

---

[2]Suppose that X and Y are Banach spaces, then X is finitely representable in Y if for all finite
dimensional subspaces E of X and all $\lambda > 1$, there is a linear map $T : E \to Y$ such that, for all
$x \in E$,

$$\lambda^{-1}\|x\|_X \le \|Tx\|_Y \le \lambda\|x\|_X.$$

A Banach space X is super-reflexive if it is reflexive and every Banach space that is finitely
representable in X is also reflexive. This concept was introduced and studied by James in [104–
106] (see [33] for additional comments).

notion of R-boundedness has proved to be a significant tool in the study of abstract multiplier operators. This condition is strictly stronger than boundedness in operator norm (besides in the Hilbert space).

The concept of R-boundedness was implicitly introduced by Bourgain in [24] and later also by Zimmermann [182]. Explicitly it is due to Berkson and Gillespie [19] and Clément et.al. [45].

We define the means

$$\|(x_1,\ldots,x_n)\|_R := \frac{1}{2^n} \sum_{\epsilon_j \in \{-1,1\}^n} \left\| \sum_{j=1}^{n} \epsilon_j x_j \right\|,$$

for $x_1,\ldots,x_n \in X$.

Definition 2.2.1. Let X and Y be Banach spaces. A subset $\mathcal{T}$ of $\mathcal{B}(X,Y)$ is called R-bounded if there is a constant $c \geq 0$ such that

$$\|(T_1 x_1,\ldots,T_n x_n)\|_R \leq c\|(x_1,\ldots,x_n)\|_R, \qquad (2.2.1)$$

for all $T_1,\ldots,T_n \in \mathcal{T}$, $x_1,\ldots,x_n \in X$, $n \in \mathbb{N}$. The least c such that (2.2.1) is satisfied is called the R-bound of $\mathcal{T}$ and is denoted $R(\mathcal{T})$.

Denote by r_j the jth Rademacher function, that is, $r_k(t) := sign(\sin(2^k \pi t))$. For $x \in X$ we denote by $r_j \otimes x$ the vector-valued function $t \rightarrow r_j(t)x$. An equivalent definition using the Rademacher functions replaces (2.2.1) by

$$\left\| \sum_{k=1}^{n} r_k \otimes T_k x_k \right\|_{L^2(0,1;Y)} \leq c \left\| \sum_{k=1}^{n} r_k \otimes x_k \right\|_{L^2(0,1;X)}. \qquad (2.2.2)$$

The Rademacher functions are an orthonormal sequence in $L^2[0,1]$. For more information, see [107] and [127, Sect. 2].

We note that Khintchine's inequality allows us to evaluate L^p-norms of Rademacher sums: For $1 \leq p < \infty$ there is a constant $C_p > 0$ such that for all $a_n \in \mathbb{C}$

$$\frac{1}{C_p} \left(\sum_n |a_n|^2\right)^{1/2} \leq \left\| \sum_n r_n a_n \right\|_{L^p[0,1]}$$

$$\leq C_p \left(\sum_n |a_n|^2\right)^{1/2}.$$

For $p = 2$ this inequality follows directly from the fact that (r_n) is an orthonormal sequence in $L^2[0,1]$ with $C_2 = 1$. For the other cases, see [127, Sect. 2].

We also note Kahane's inequality: For any Banach space X and $1 \leq p,q \leq \infty$ there is a positive constant $C(p,q,X)$ such that

$$\left\| \sum_{j=1}^{N} r_j \otimes x_j \right\|_{L_p(\Omega;X)} \leq C(p,q,X) \left\| \sum_{j=1}^{N} r_j \otimes x_j \right\|_{L_q(\Omega;X)}.$$

Remark 2.2.2. (i) A uniformly bounded family of operators is not necessarily R-bounded, as can be seen by considering the set of translation operator $T_n f(\cdot) = f(\cdot - n)$, on $X = L^p$, with $1 \leq p < \infty$ and $p \neq 2$ (see [127] for details).

(ii) If $\mathcal{T} = \{T^k\}_{k=1}^{\infty} \subset \mathcal{B}(X,Y)$ is a countable sequence of operators, then it is sufficient to verify the inequality (2.2.2) for all truncated sequences $\{T^k\}_{k=1}^{n}$ of the first n members of the sequence.[3]

Example 2.2.3. Let X and Y be Banach spaces.

(a) Let $J \subset \mathbb{R}$ be an interval and $M : J \to \mathcal{B}(X,Y)$ have an integrable derivative. Then $\{M(t) : t \in J\}$ is R-bounded.

(b) If $M : [a,b] \to \mathcal{B}(X,Y)$ is of bounded variation, then $\mathcal{N} := \{M(t) : t \in [a,b]\}$ is R-bounded with

$$R(\mathcal{N}) \leq C(\|M(a)\| + Var(M)).$$

(c) Let $p \in (1,\infty)$ and let Γ be a measure space equipped with a σ-finite measure. Let $\Phi \subset L^{\infty}(\Gamma, \mathbb{C})$ be uniformly bounded, then

$$R\left(\left\{ m_{\phi} : L^p(\Gamma, X) \to L^p(\Gamma, X) : f \to \phi f \right\}_{\phi \in \Phi}\right) \leq 2 \sup_{\phi \in \Phi} |\phi|_{L^{\infty}(\Gamma, \mathbb{C})}.$$

(d) For $T \in \mathcal{B}(X,Y)$ define the operator $(T^{\natural} f)(\lambda) := T(f(\lambda))$, $f \in L^p(\Gamma, X)$, $\lambda \in \Gamma$, $1 \leq p < \infty$. Then if $\mathcal{T} \subset \mathcal{B}(X,Y)$ is R-bounded, the collection

$$\mathcal{T}^{\natural} = \{T^{\natural} : T \in \mathcal{T}\} \subset \mathcal{B}(L^p(\Gamma, X), L^p(\Gamma, Y))$$

is also R-bounded.

(e) Let (Ω, μ) be a σ-additive measure space and $1 < p < \infty$. Let $\mathcal{T} \subset \mathcal{B}(L^p((\Omega, \mu))$ and assume that the operators in $\mathcal{T}$ are dominated by a positive operator $T \in \mathcal{B}(L^p(\Omega, \mu))$ in the sense that

$$|Sx| \leq T|x|, \quad \text{for all } S \in \mathcal{T}, x \in L^p(\Omega, \mu).$$

Then $\mathcal{T}$ is R-bounded.

---

[3]It is clear that the R-boundedness of the countable set $\mathcal{T}$ is independent of the order in which we enumerate its element. Thus it is interesting to note that, given any enumeration, the subset of n first members of the sequence are fully representative of all finite subsets of $\mathcal{T}$ in view of R-boundedness.

(f) Let $X = L^p(\Omega, \mu)$ with $1 < p < \infty$. Then $\mathcal{T} \subset \mathcal{B}(X)$ is R-bounded if and only if

$$\mathcal{T}' = \{T' : T \in \mathcal{T}\} \subset \mathcal{B}(X')$$

is R-bounded and $R(\mathcal{T}') = R(\mathcal{T})$.

(g) Let $f \in \mathcal{S}(\mathbb{R}^n, \mathcal{B}(X, Y))$. Then the range $f(\mathbb{R}^n) \subset \mathcal{B}(X, Y)$ is R-bounded.

(h) Let G be an index set and let $D \subset \mathbb{R}^n$ be open and consider a family $\mathcal{T} = \{T_\mu : \mu \in G\} \subset \mathcal{B}(L^p(D, \mathbb{C}^m))$ of kernel operators

$$T_\mu f(x) = \int_D k_\mu(x, y) f(y) dy, \ x \in D, \ f \in L^p(D, \mathbb{C}^m),$$

which are dominated by a kernel k_0, i.e.,

$$|k_\mu(x, y)| \le k_0(x, y) \text{ for a.e. } x, y \in D, \text{ and all } \mu \in G.$$

Then $\mathcal{T} \subset \mathcal{B}(L^p(D, \mathbb{C}^m))$ is R-bounded provided T_0 is bounded in $L^p(D, \mathbb{C}^m)$.

(i) Let $\mathcal{G} \subset \mathbb{C}$ be a simply connected Jordan region such that $\mathbb{C} \backslash \mathcal{G}$ has interior points. Let $F \in L^\infty(\overline{\mathcal{G}}; \mathcal{B}(X))$ be analytic in $\mathcal{G}$ and $F(\partial \mathcal{G})$ be R-bounded. Then $F(\overline{\mathcal{G}})$ is also R-bounded.

The proof of Example 2.2.3 (a) [resp. (b)] is in [180] (resp. [174]). Examples 2.2.3 (c), (e), and (f) are taken from [127]. The proof of (g) [resp. (d)] is in [66] (resp. [174]). The proof of (h) is in [65]. (i) is taken from [22].

Remark 2.2.4. If $\mathcal{T}$ is an R-bounded subset of $\mathcal{B}(X, Y)$ for arbitrary Banach spaces X and Y, then

$$\mathcal{T}'' = \{T'' : T \in \mathcal{T}\} \subset \mathcal{B}(X'', Y'')$$

is R-bounded, but in order to conclude that $\mathcal{T}' = \{T' : T \in \mathcal{T}\}$ is R-bounded in $\mathcal{B}(X', Y')$ we have to assume that X and Y are of nontrivial type (see [127]).

R-boundedness has a number of nice permanence properties, some of these are summarized in the following two propositions.

Proposition 2.2.5. *Let X and Y be Banach spaces. We have the following properties:*

(a) *Any finite family $\mathcal{T} \subset \mathcal{B}(X, Y)$ is R-bounded.*

(b) *If $\mathcal{T} \subset \mathcal{B}(X, Y)$ is R-bounded, then it is uniformly bounded, with*

$$\sup\{\|T\| : T \in \mathcal{T}\} \le R(\mathcal{T}).$$

(c) *The definition of R-boundedness is independent of $p \in [1, \infty)$.*

(d) *When X and Y are Hilbert spaces, $\mathcal{T} \subset \mathcal{B}(X, Y)$ is R-bounded if and only if $\mathcal{T}$ is uniformly bounded.*

(e) *Let $\mathcal{D}, \mathcal{T} \subset \mathcal{B}(X, Y)$ be R-bounded sets, then*

$$\mathcal{D} \pm \mathcal{T} := \{S \pm T : S \in \mathcal{D}, T \in \mathcal{T}\}$$

is R- bounded and $R(\mathcal{D} \pm \mathcal{T}) \leq R(\mathcal{D}) + R(\mathcal{T})$.

(f) *Let $\mathcal{T} \subset \mathcal{B}(X, Y)$ and $\mathcal{D} \subset \mathcal{B}(Y, Z)$ be R-bounded sets, then*

$$\mathcal{D} \cdot \mathcal{T} := \{S \cdot T : S \in \mathcal{D}, T \in \mathcal{T}\} \subset \mathcal{B}(X, Z)$$

is R-bounded and $R(\mathcal{D} \cdot \mathcal{T}) \leq R(\mathcal{D}) \cdot R(\mathcal{T})$.

(g) *A subset $\mathcal{T} \subset \mathcal{B}(X)$ of the form $\mathcal{T} = \{\lambda I : \lambda \in \mathcal{U}\}$ is R-bounded, whenever $\mathcal{U} \subset \mathbb{C}$ is bounded.*

(h) *Let $\mathcal{G} \subset \mathbb{C}$ be open and let $\lambda \in \mathcal{G} \to N(\lambda) \in \mathcal{B}(X)$ be an analytic function. Then the set*

$$\{N(\lambda) : \lambda \in K\}$$

is R-bounded for every compact set $K \subset \mathcal{G}$.

(i) *Let $\mathcal{T} \in \mathcal{B}(X)$ be an R-bounded set, then the strong closure of $\mathcal{T}$ is also R-bounded.*

(j) *Let $\mathcal{T} \in \mathcal{B}(X)$ be an R-bounded set, then the convex hull of $\mathcal{T}$*

$$co(\mathcal{T}) = \left\{ \sum_{j=1}^{n} \lambda_j T_j : T_j \in \mathcal{T}, \lambda_j \in \mathbb{C}, \sum_{j=1}^{n} |\lambda_j| \leq 1, n \in \mathbb{N} \right\}$$

and the absolute convex hull of $\mathcal{T}$

$$abco(\mathcal{T}) = \left\{ \sum_{j=1}^{n} \lambda_j T_j : T_j \in \mathcal{T}, \lambda_j \in \mathbb{C}, \sum_{j=1}^{n} |\lambda_j| = 1, n \in \mathbb{N} \right\}$$

are R-bounded, and $R(co(\mathcal{T})) \leq 2R(\mathcal{T})$, $R(abco(\mathcal{T})) \leq 2R(\mathcal{T})$.

Proposition 2.2.6. *Let X and Y be Banach spaces. We have the following properties:*

(a) *Let G be an index set and assume that*

$$T(\mu) = \sum_{n=1}^{\infty} T_n(\mu), \ \mu \in G,$$

converges in the strong operator topology of $\mathcal{B}(X, Y)$ for all $\mu \in G$. Then

$$R\left(\left\{T(\mu) : \mu \in G\right\}\right) \leq \sum_{n=1}^{\infty} R\left(\left\{T_n(\mu) : \mu \in G\right\}\right).$$

(b) *Let T be an R-bounded set in $\mathcal{B}(X, Y)$. For every strongly measurable $N : \Gamma \to \mathcal{B}(X, Y)$ on a σ-finite measure space Γ with values in T and every $h \in L^1(\Gamma, \mu)$ we define an operator $T_{N,h} \in \mathcal{B}(X, Y)$ by*

$$T_{N,h}x = \int_\Gamma h(\omega)N(\omega)x d\mu(\omega), \ x \in X.$$

Then $\mathfrak{F} := \{T_{N,h} : ||h||_{L^1} \leq 1, \ N \ as above\}$ is R-bounded and $R(\mathfrak{F}) \leq 2R(T)$.

(c) *Let $\mathcal{U}$ be a subset of $\mathbb{C}$. Assume that $T : \Gamma \times \mathcal{U} \to \mathcal{B}(X, Y)$ is such that $T(\cdot, \lambda)$ is μ-integrable in $\mathcal{B}(X, Y)$ for all $\lambda \in \mathcal{U}$ and that $T(\Gamma \times \mathcal{U})$ is R-bounded. Then*

$$\left\{\int_\Gamma T(\omega, \lambda)d\mu(\omega) : \lambda \in \mathcal{U}\right\}$$

is R-bounded.

(d) *Let $T \subset \mathcal{B}(X, Y)$ be an R-bounded set. If a collection*

$$\mathcal{N} = \{M_n : n \in \mathbb{N}\}$$

has the property that $C := \sup\{||M_n||_T : n \in \mathbb{N}\} < \infty$, where $|| \cdot ||_T$ denotes the Minkowski functional[4] of $abco(T)$, then $\mathcal{N}$ is also R-bounded and $R(\mathcal{N}) \leq 4CR(\mathcal{N})$.

(e) *Let $D \subset \mathbb{R}^n$ be open and $1 < p < \infty$. Suppose $T \subset \mathcal{B}(L^p(D, X), L^p(D, Y))$ is a family of kernel operators in the sense that*

$$Kf(x) = \int_D k(x, x')f(x')dx', \ x \in D, \ f \in L^p(D, X),$$

for each $K \in T$, where the kernels $k : D \times D \to \mathcal{B}(X, Y)$ are measurable and such that

$$R\left(\left\{k(x, x') : k \in T\right\}\right) \leq k_0(x, x'), \ x, x' \in D,$$

and the operator K_0 with scalar kernel k_0 is bounded in $L^p(D)$. Then the family $T \subset \mathcal{B}(L^p(D, X), L^p(D, Y))$ is R-bounded and $R(T) \leq ||K_0||_{L^p(D)}$.

(f) *Let $T \subset \mathcal{B}(X)$ be an R-bounded set and $C > 0$, $q \in [0, 1)$. Then for $\mathcal{A} := \{a \in l^\infty : \forall n \in \mathbb{N}, |a_n| \leq C(q/R(T))^n\}$ the set*

---

[4]For a bounded collection $T \subset \mathcal{B}(X, Y)$ we denote the Minkowski functional of $abco(T)$ by $|| \cdot ||_T : \mathcal{B}(X, Y) \to [0, \infty]$, $T \to ||T||_T = \inf\{t > 0 : T \in t \cdot abco(T)\}$ (see [174]).

$$\left\{ \sum_{n=1}^{\infty} a_n T^n : a \in \mathcal{A}, T \in \mathcal{T} \right\}$$

is R-bounded.

For a detailed proof of Propositions 2.2.5 and 2.2.6 we refer to the monograph of Denk–Hieber and Prüss [65] (see also [22, 127, 174, 180]).

Proposition 2.2.7 ([22]). *Let A be a closed and densely defined operator in X:*

(a) *If $\{\lambda R(\lambda, A) : \lambda \in \Sigma_{\pi/2}\}$ is R-bounded, then there is $\delta > \pi/2$ such that $\{\lambda R(\lambda, A) : \lambda \in \Sigma_{\delta}\}$ is R-bounded.*
(b) *If $\mathcal{G} \subset \rho(A)$ is compact, then $\{R(\lambda, A) : \lambda \in \mathcal{G}\}$ is R-bounded.*

Let $\mathcal{E}$ be the set of all entire $\mathbb{C}$-valued functions. For all $f \in \mathcal{E}$ we define

$$f(T) := \int_{\Gamma_f} f(\lambda) R(\lambda, T) d\lambda,$$

where Γ_f is an arbitrary path in $\rho(T)$ around $\sigma(T)$. Note that this definition is independent of the chosen path Γ_f.

Proposition 2.2.8 ([22]). *Let $\mathcal{G} \subset \rho(T)$ and $g : \mathcal{G} \to \mathbb{C} \backslash \{0\}$ be continuous such that $\{g(\lambda) R(\lambda, T) : \lambda \in \mathcal{G}\}$ is R-bounded. Let $\mathcal{F} \subset \mathcal{E}$ be such that $\Gamma_f \subset \mathcal{G}$ for all $f \in \mathcal{F}$ and*

$$\sup_{f \in \mathcal{F}} \int_{\Gamma_f} |f(\lambda) g(\lambda)^{-1}| |d\lambda| < \infty.$$

Then the set $\{f(T) : f \in \mathcal{F}\}$ is R-bounded.

2.3 Maximal L^p-Regularity

In the last 15 years, a lot of progress has been made on the subject of maximal regularity. The problem of (parabolic) maximal L^p-regularity can be stated as follows:

Let A be a linear closed densely defined operator on a Banach space X, with domain $D(A)$.

Let $p \in (1, \infty)$. We say that there is maximal L^p-regularity on the interval I (with $I = [0, T]$ or $I = [0, \infty)$) for the Cauchy problem

$$\begin{cases} u'(t) = Au(t) + f(t), \, t \in I, \\ \\ u(0) = 0, \end{cases} \tag{2.3.1}$$

if for every $f \in L^p(I;X)$ there exists one and only one $u \in L^p(I;D(A)) \cap W^{1,p}(I;X)$ satisfying the Cauchy problem (2.3.1). In this case $Au \in L^p(I;X)$ as well. In other words, maximal L^p-regularity means that each term in the abstract Cauchy problem is well defined and has the same regularity.[5]

From the closed graph theorem it follows that if there is maximal L^p-regularity, then there is $C \in \mathbb{R}^+$ such that

$$\|u\|_{L^p} + \|u'\|_{L^p} + \|Au\|_{L^p} \le C \|f\|_{L^p}. \tag{2.3.2}$$

Sobolevskii [169] showed that if (2.3.1) has maximal L^p-regularity for some $1 < p < \infty$, then it has maximal L^q-regularity for all $1 < q < \infty$.

As a consequence of maximal L^p-regularity one can obtain estimates of the resolvent operator $(\lambda - A)^{-1}$ (see [68, Theorem 2.1]).

Theorem 2.3.1. *Let X be a complex Banach space and A a linear closed densely defined operator in X such that there is maximal L^p-regularity on the interval $[0, \infty)$ for (2.3.1). Then*

$$\left\{\lambda \in \mathbb{C} : Re(\lambda) \ge 0\right\} \subset \rho(A)$$

and there is $C > 0$ such that

$$Re(\lambda) \ge 0 \Rightarrow \|(\lambda - A)^{-1}\| \le \frac{C}{1 + |\lambda|}.$$

Remark 2.3.2. If there is maximal L^p-regularity on the interval $[0, T]$ for (2.3.1), then there is $\delta \in \mathbb{R}^+$ such that

$$\left\{\lambda \in \mathbb{C} : Re(\lambda) \ge \delta\right\} \subset \rho(A)$$

and there is $C > 0$ such that

$$Re(\lambda) \ge \delta \Rightarrow \|(\lambda - A)^{-1}\| \le \frac{C}{1 + |\lambda|}.$$

In particular, A generates an analytic semigroup. It means that maximal L^p-regularity can only occur in parabolic problems [68].

---

[5]The first positive result on maximal L^p-regularity was obtained by Ladyzhenskaya, Solonnikov, and Ural'tseva [128], where $X = L^p(G)$, $G \subset \mathbb{R}^n$ being a bounded domain with smooth boundary, A a strongly elliptic second- order differential operator with continuous coefficients, and $1 < p < \infty$. The first abstract result was obtained by de Simon [62] for Hilbert spaces. Specifically, let H be a Hilbert space and A be the generator of an analytic semigroup. Then (2.3.1) has maximal L^p-regularity on $[0, \infty)$. De Simon's proof employ Plancherel's theorem which is known to be valid only in the Hilbert space case (see [65, 68]).

Remark 2.3.3 ([68]). (a) If there is maximal L^p-regularity on the interval $[0, \infty)$ for (2.3.1), then for each $T \in \mathbb{R}^+$, the problem (2.3.1) has maximal L^p-regularity on the interval $[0, T]$.

(b) If there is maximal L^p-regularity on the interval $[0, T]$ for (2.3.1) and the semigroup generated by A has negative exponential type, then there is maximal L^p-regularity on $[0, \infty)$.

(c) There is a Banach space X and an operator A such that (2.3.1) has maximal L^p-regularity on $[0, T]$ for all $T \in \mathbb{R}^+$ but (2.3.1) does not have maximal L^p-regularity on $[0, \infty)$ (see [132]).

(d) If there exists $T_0 \in \mathbb{R}^+$ such that there is maximal L^p-regularity on the interval $[0, T_0]$ for (2.3.1), then for all $T \in \mathbb{R}^+$ there is maximal L^p-regularity on $[0, T]$. That is, maximal L^p-regularity does not depend on the compact interval on which we consider (2.3.1).

Theorem 2.3.4 ([68]). *Let X be a Banach space and A a closed linear operator in X that generates a strongly continuous analytic semigroup. If there exists $\tilde{p} \in (1, \infty)$ such that there is maximal $L^{\tilde{p}}$-regularity on $[0, T]$ for (2.3.1), then for each $p \in (1, \infty)$ there is maximal L^p-regularity.*

Theorem 2.3.5 ([68]). *Let Y be a Banach space and A a closed linear operator in Y that generates a strongly continuous analytic semigroup with negative exponential type. Choose $\theta \in (0, 1)$, $p \in (1, \infty)$ and put $X = (Y, D(A))_{\theta, p}$, then there is maximal L^p-regularity on $[0, \infty)$ for (2.3.1) in the space X.*

Theorem 2.3.6 ([68]). *Let X be a UMD space and A a closed linear densely defined operator in X such that*

(a) *$-A$ is positive, that is, $(-\infty, 0] \subset \rho(-A)$ and $\|(\lambda + A)^{-1}\| \leq C(1 + |\lambda|)^{-1}$ for every $\lambda \in (-\infty, 0]$.*

(b) *The operator $(-A)^{it}$ is bounded and there are $\theta \in [0, \pi/2)$ and $c \in \mathbb{R}^+$ such that for each $t \in \mathbb{R}$, $\|(-A)^{it}\| \leq Ce^{\theta|t|}$.*

Then for each $p \in (1, \infty)$ there is maximal L^p-regularity on $[0, T]$ for (2.3.1).

Theorem 2.3.7 ([112]). *Let X be a Banach space with an unconditional basis.[6] Assume that, for each analytic semigroup $(e^{tA})_{t \in \mathbb{R}^+}$, the Cauchy problem (2.3.1) has maximal L^p-regularity on $[0, T]$. Then X is isomorphic to l_2.*

The maximal regularity, from theoretical point of view, has been approached in different ways. For an overview, see the recent survey of Monniaux [147] and references therein.

One way to treat the problem of maximal regularity is to apply the Fourier transform to (2.3.1). The problem is now to decide whether

$$M(t) = A(isI - A)^{-1}, s \in \mathbb{R},$$

---

[6]Let X be a Banach space, $(x_k)_{k \in \mathbb{N}} \subset X$ is called a Schauder basis if, for every $x \in X$, there is a unique sequence $(a_k)_{k \in \mathbb{N}} \subset \mathbb{C}$ such that $x = \sum_{k=1}^{\infty} a_k x_k$. It is called an unconditional basis if the series converges unconditionally.

is a Fourier multiplier. This has been studied by Weis in [180] who gave an equivalent property to maximal regularity of A in terms of bounds of the resolvent of A. The key tools are operator-valued Fourier multiplier theorems established independently by Amann and Weis (see [4, 179, 180]). The result is as follows:

Theorem 2.3.8. *Let X be a UMD space and let $(e^{tA})_{t \geq 0}$ be a bounded analytic semigroup on X (see Sect. 1.5). Then the following are equivalent:*

(a) *Equation (2.3.1) has maximal regularity.*
(b) *$\{\lambda(\lambda - A)^{-1} : \lambda \in i\mathbb{R}, \lambda \neq 0\}$ is R-bounded.*
(c) *$\{e^{tA}, tAe^{tA} : t \geq 0\}$ is R-bounded.*

After these pioneering works, the theory of differential operator equations in Banach-valued function spaces improved significantly. Many researchers applied them in the investigation of different classes of equations especially in maximal L^p (Lebesgue) and C^s (Hölder) regularity for parabolic and elliptic differential operator equations. As to literature, there has been a substantial amount of work, as one can see, for example, in Amann [7], Denk–Hieber and Prüss [65], Clément–Londen–Simonett [46], the survey by Arendt [10], and the bibliography therein.

The property of maximal regularity has many important applications to evolution equations. It is an important tool in the study of the following problems: existence and uniqueness of solutions of nonautonomous evolution equations; existence and uniqueness of solutions of quasilinear and nonlinear partial differential equations; stability theory for evolution equations, or in others words construction of central manifolds, and detection of bifurcations; maximal regularity of solutions of elliptic differential equations; existence and uniqueness of solutions of Volterra integral equations; and uniqueness of mild solutions of the Navier–Stokes equations.

In these applications, a maximal regularity is frequently used to reduce, via a fixed-point argument, a nonautonomous (resp. nonlinear) problem to an autonomous (resp. a linear) problem. In some cases, maximal regularity is needed to apply an implicit function theorem.

2.4 Vector-Valued Fourier Multipliers: Blunck's Theorem

In this section we collect some general operator-valued multiplier theorems from the literature. Here UMD spaces enter in the history on maximal regularity. Fourier multipliers on vector-valued functions spaces are needed to establish existence and uniqueness as well as regularity of differential equations in Banach spaces and thus also for partial differential equations.

Recall that the Schwartz class $\mathcal{S}(\mathbb{R}^n; X)$ of rapidly decreasing smooth functions from $\mathbb{R}^n$ into X is norm dense in $L^p(\mathbb{R}^n; X)$ for $1 \leq p < \infty$. The Fourier transform $\mathcal{F} : \mathcal{S}(\mathbb{R}^n; X) \to \mathcal{S}(\mathbb{R}^n; X)$ defined by

$$\mathcal{F}f(\xi) \equiv \hat{f}(\xi) := \int_{\mathbb{R}^n} e^{-i\xi \cdot \eta} f(\eta) d\eta, \tag{2.4.1}$$

is a bijection whose inverse is given by

$$\mathcal{F}^{-1} f(\xi) \equiv \check{f}(\xi) := (2\pi)^{-n} \int_{\mathbb{R}^n} e^{i\xi \cdot \eta} f(\eta) d\eta, \qquad (2.4.2)$$

where $f \in \mathcal{S}(\mathbb{R}^n; X)$ and $\xi \in \mathbb{R}^n$. Note that the formula in (2.4.1) [resp. (2.4.2)] defines a mapping $\mathcal{F}$ (resp. $\mathcal{F}^{-1}$) in $\mathcal{B}(L^1(\mathbb{R}^n, X), L^\infty(\mathbb{R}^n, X))$.

Definition 2.4.1 ([180]). We say that a function $M : \mathbb{R}\backslash\{0\} \to \mathcal{B}(X, Y)$ is a Fourier multiplier on $L^p(\mathbb{R}, X)$ if the expressions

$$T_M f = \mathcal{F}^{-1}[M(\cdot)\mathcal{F}f(\cdot)], \ f \in \mathcal{S}(\mathbb{R}; X), \qquad (2.4.3)$$

are well defined and T_M extends to a bounded operator

$$T_M : L^p(\mathbb{R}, X) \to L^p(\mathbb{R}, Y).$$

J. Schwartz proved the following well-known result.[7]

Theorem 2.4.2. *Let X be a Hilbert space. Assume that for the function $M \in C^1(\mathbb{R}\backslash\{0\}, \mathcal{B}(X))$, the sets*

$$\{M(t) : t \in \mathbb{R}\backslash\{0\}\} \quad and \quad \{tM'(t) : \mathbb{R}\backslash\{0\}\} \qquad (2.4.4)$$

are bounded in $\mathcal{B}(X)$. Then the Fourier multiplier operator (2.4.3) extends to a bounded operator T_M on $L^p(\mathbb{R}, X)$, for $1 < p < \infty$.

Pisier observed that the converse is true: if all M satisfying (2.4.4) are Fourier multiplier on $L^2(\mathbb{R}, X)$, then X is isomorphic to a Hilbert space. Therefore, additional hypotheses are needed to obtain multiplier theorems in more general spaces. In the context of the Mikhlin-multiplier theorem besides the UMD property for X and Y the additional condition can be expressed in terms of R-boundedness.

We have the following Fourier multiplier theorem due to Weis [180].

Theorem 2.4.3. *Let X and Y be UMD spaces. Assume that for*

$$M \in C^1(\mathbb{R}\backslash\{0\}, \mathcal{B}(X, Y))$$

the sets in (2.4.4) are R-bounded in $\mathcal{B}(X, Y)$. Then the Fourier multiplier operator (2.4.3) extends to a bounded operator $T_M : L^p(\mathbb{R}, X) \to L^p(\mathbb{R}, Y)$, for $1 < p < \infty$.

Note that if $X = Y$ is a Hilbert space, this theorem reduces to Theorem 2.4.2.

---

[7]A proof of J. Schwartz's result (Theorem 2.4.2) using the Calderon–Zygmund method can be found in [18].

The preceding result also holds in the periodic case by a result of Arendt and Bu [8]. We consider the Banach space $L^p(0, 2\pi; X)$ of all X-valued Bochner measurable functions f on $[0, 2\pi]$ with the norm

$$\|f\|_p := \left(\int_0^{2\pi} \|f(t)\|^p dt \right)^{1/p}, \quad \text{where } 1 \le p < \infty.$$

For $f \in L^p(0, 2\pi; X)$ we denote

$$\hat{f}(k) := \frac{1}{2\pi} \int_0^{2\pi} e^{-ikt} f(t) dt,$$

the kth Fourier coefficient of f, where $k \in \mathbb{Z}$.

Definition 2.4.4 ([8]). Let X and Y be two Banach spaces, $1 \le p < \infty$; we say that a sequence $\{M_k\}_{k\in\mathbb{Z}} \subset \mathcal{B}(X, Y)$ is an L^p-multiplier if for each $f \in L^p(0, 2\pi; X)$ there is $g \in L^p(0, 2\pi; Y)$ such that

$$\hat{g}(k) = M_k \hat{f}(k),$$

for all $k \in \mathbb{Z}$.

In this case there is a unique operator $M \in \mathcal{B}(L^p(0, 2\pi; X), L^p(0, 2\pi; Y))$ such that

$$\widehat{Mf}(k) = M_k \hat{f}(k), \ k \in \mathbb{Z}.$$

for all $f \in L^p(0, 2\pi; X)$. We call M the operator associated with $\{M_k\}_{k\in\mathbb{Z}}$. On the other hand, we have

$$\{M_k : k \in \mathbb{Z}\}$$

as R-bounded (see [8, Proposition 1.11]). This indicates that R-boundedness arises naturally in the context of multiplier theorems.

Remark 2.4.5. On a Hilbert space H each bounded sequence $\{M_k\}_{k\in\mathbb{Z}} \subset \mathcal{B}(H)$ is an L^2-multiplier. This follows from the fact that the Fourier transform given by

$$f \in L^p(0, 2\pi; H) \to \{\hat{f}(k)\}_{k\in\mathbb{Z}} \in l^2(\mathbb{Z}; H)$$

is an isometric isomorphism.

Remark 2.4.6. If X is a UMD space, then

$$R\left(\sum_{k=-N}^{N} e_k \otimes x_k \right) = \sum_{k=0}^{N} e_k \otimes x_k$$

defines an L^p-multiplier for $1 < p < \infty$, which is called the Riesz projection (see Chap. 7 for notations and [8, 32]).

Remark 2.4.7. Let X be a *UMD* space and $1 < p < \infty$. Define the projections P_l on $L^p(0, 2\pi; X)$ by

$$P_l(\sum_{k \in \mathbb{Z}} e_k \otimes x_k) = \sum_{k \geq l} e_k \otimes x_k.$$

Then the set $\{P_l : l \in \mathbb{Z}\}$ is R-bounded in $\mathcal{B}(L^p(0, 2\pi; X))$.

The following multiplier theorem has been obtained in [8] which is a discrete analog of the operator-valued version of Mikhlin's theorem due to Weis [180].

Theorem 2.4.8 (Marcinkiewicz operator-valued multiplier theorem). *Let X and Y be UMD spaces and let $\{M_k\}_{k \in \mathbb{Z}} \subset \mathcal{B}(X, Y)$ be a sequence. If the families $\{M_k\}_{k \in \mathbb{Z}}$ and $\{k(M_{k+1} - M_k)\}_{k \in \mathbb{Z}}$ are R-bounded, then $\{M_k\}_{k \in \mathbb{Z}}$ is an L^p-multiplier for $1 \leq p < \infty$.*

In discrete time, the operator-valued Fourier multiplier theorem is due to Blunck [22, Theorem 1.3], which plays a key role in our analysis of maximal regularity for difference equations.

Theorem 2.4.9. *Let $p \in (1, \infty)$ and let X be a UMD space. Let $M : (-\pi, \pi) \setminus \{0\} \to \mathcal{B}(X)$ be a differentiable function such that the set*

$$\{M(t), (e^{it} - 1)(e^{it} + 1)M'(t) : t \in (-\pi, \pi) \setminus \{0\}\}$$

is R-bounded. Then there is an operator $T_M \in \mathcal{B}(l_p(\mathbb{Z}; X))$ such that

$$\mathcal{F}(T_M f)(e^{it}) = M(t)\mathcal{F}f(e^{it}), \quad t \in (-\pi, \pi) \setminus \{0\}, \ \mathcal{F}f \in L^\infty(\mathbb{T}; X)$$

of compact support.

Remark 2.4.10. Recall that the discrete-time Fourier transform considered in Theorem 2.4.9 corresponds to the Z-transform around the unit circle in the complex plane.

An application of Blunck's theorem to difference equations is in their infancy. Most contributions in this area are due to Portal [160–162] and the works of the authors [53–55] and collaborators.

2.5 Comments

Maximal L^p-Regularity for $p \in \{1, \infty\}$

Besides Hilbert spaces $L^\infty([0, T])$ has the maximal regularity property for the simple reason that every generator of a strongly continuous semigroup is already bounded by the famous result of Lotz [137].

The main results on the maximal L^p-regularity with $p = \infty$ and $p = 1$ were obtained by Baillon [14] and Guerre-Delabrière [90], respectively.

Theorem 2.5.1. *Let X be a separable Banach space and let $(A, D(A))$ be the generator of an analytic semigroup. Then the following conditions are equivalent:*

(a) *X does not contain a copy of c_0.*
(b) *If (2.3.1) has maximal L^∞-regularity on $[0, T]$, then A is bounded.*

Theorem 2.5.2. *Let X be a Banach space and let $(A, D(A))$ be the generator of an analytic semigroup. Then the following conditions are equivalent:*

(a) *X does not contain a complemented copy of l_1.*
(b) *If (2.3.1) has maximal L^1-regularity on $[0, T]$, then A is bounded.*

The following two results are due to Kalton and Portal [111].

Theorem 2.5.3. *Let $-A$ be the generator of a bounded analytic semigroup. The following conditions are equivalent:*

(a) *Equation (2.3.1) has maximal L^1-regularity on $[0, \infty)$.*
(b) *There is a constant $C > 0$ so that*

$$\int_0^\infty ||Ae^{-tA}x||dt \leq C||x||, \quad x \in X.$$

Theorem 2.5.4. *Let $-A$ be the generator of a bounded analytic semigroup. The following conditions are equivalent:*

(a) *Equation (2.3.1) has maximal L^∞-regularity on $[0, \infty)$.*
(b) *There is a constant $C > 0$ so that*

$$||x|| \leq C \sup_{t>0} ||tAe^{-tA}x|| + \limsup_{t \to \infty} ||e^{-tA}x||, \quad x \in X.$$

Remark 2.5.5. If A has a dense range, then we have $\lim_{t \to \infty} e^{-tA}x = 0$ for every $x \in X$ and we can drop the last term in (b) (see [111]).

Fackler [79] gave recently a more explicit proof of a result by Kalton and Lancien stating that on each Banach space with an unconditional basis not isomorphic to a Hilbert space there exists a generator A of a holomorphic semigroup which does not have maximal regularity. In particular, Fackler has shown that there always exists a Schauder basis (f_m) such that A can be chosen of the form $A(\sum_{m=1}^{\infty} a_m f_m) = \sum_{m=1}^{\infty} 2^m a_m f_m$. Moreover, he proved that maximal regularity does not extrapolate: He constructed consistent holomorphic semigroups on $L^p(\mathbb{R})$ for $p \in (1, \infty)$ which have maximal regularity if and only if $p = 2$.

More About UMD Spaces

X is a UMD space if and only if it is ξ-convex, that is, there is a biconvex function $\xi : X \times X \to \mathbb{R}$ (i.e., if both $\xi(\cdot, y)$ and $\xi(x, \cdot)$ are convex on X for all $x, y \in X$) such that $\xi(0, 0) > 0$ and for every $(x, y) \in X \times X$ such that $||x|| = ||y|| = 1$, we have $\xi(x, y) \leq ||x + y||$ (see [33, 161]). ξ-convexity is preserved by Banach space isomorphisms; closed subspaces and cartesian products of ξ-convex spaces are ξ-convex. If (X, Y) is an interpolation couple of ξ-convex spaces, then the complex interpolation spaces $[X, Y]_\theta$ and the real interpolation spaces $(X, Y)_{\theta, p}$ with $1 < p < \infty$, are ξ-convex spaces. For more information about ξ-convex spaces see Rubio de Francia's survey article [165].

There are many important statements in vector-valued harmonic analysis and probability theory that are equivalent to the UMD property. We refer the interested reader to [31]. The significance of the UMD property for vector-valued multiplier theorems was recognized in [25].

Miscellaneous on R-Boundedness

R-boundedness generalizes the notion of square function estimates in L^q [161]. More precisely consider $T \in \mathcal{B}(L^q)$ for some $1 < q < \infty$, T is R-bounded if and only if $\exists C > 0$, $\forall n \in \mathbb{N}$, $\forall T_1, \ldots, T_n \in T$, $\forall x_1, \ldots, x_n \in L^q$

$$\left\|\left(\sum_{j=1}^n |T_j x_j|^2\right)^{\frac{1}{2}}\right\|_q \leq C \left\|\left(\sum_{j=1}^n |x_j|^2\right)^{\frac{1}{2}}\right\|_q.$$

The notion of R-boundedness was named "Riesz property" in [19], "Randomized boundedness" in [45], and it is also known as "Rademacher boundedness." Some interesting surveys on R-boundedness can be found in [34, 100, 181].

A nice discussion of the Rademacher functions can be found in Kac's book [109].

Proposition 2.5.6 ([127] (Kahane's contraction principle)). *For all $(a_n) \subseteq \mathbb{C}$ with $|a_n| \leq 1$,*

$$\left\|\sum_n r_n \otimes a_n x_n\right\|_{L^p([0,1], X)} \leq 2 \left\|\sum_n r_n \otimes x_n\right\|_{L^p([0,1], X)}.$$

We recall that a Banach space X is of type $1 \leq p \leq 2$ if there is $C > 0$ such that for $x_1, x_2, \cdots, x_n \in X$ we have

$$\left\| \sum_{k=1}^{n} r_k \otimes x_k \right\|_{L^2([0,1],X)} \leq C \left(\sum_{k=1}^{n} \|x_k\|^p \right)^{1/p}.$$

X is of cotype $2 \leq q \leq \infty$ if there is $C' > 0$ such that for $x_1, x_2, \cdots, x_n \in X$ we have

$$\left(\sum_{k=1}^{n} \|x_k\|^q \right)^{1/q} \leq C' \left\| \sum_{k=1}^{n} r_k \otimes x_k \right\|_{L^2([0,1],X)}$$

(with the usual modification if $q = \infty$) [158] (see also [133]).

In contrast with the Proposition 2.2.5 (d) we have the following more general proposition.

Proposition 2.5.7 ([8]). *Let X and Y be two Banach spaces. Then the following assertions are equivalent:*

 (i) *X is of cotype 2 and Y is of type 2.*
(ii) *Each bounded subset in $\mathcal{B}(X, Y)$ is R-bounded.*

One of the main features of R-bounded sets of operators is the following result, due to Clément et al. [45].

Proposition 2.5.8. *Let $(\nabla_n)_{n \in \mathbb{Z}} \subset \mathcal{B}(X)$ be a unconditional Schauder decomposition of X and let $T \subset \mathcal{B}(X)$ be R-bounded with $T \nabla_n = \nabla_n T$ for all $n \in \mathbb{Z}$. Then each sequence $(T_n)_{n \in \mathbb{Z}} \subset T$ induces via*

$$x \to \sum_{n \in \mathbb{Z}} T \nabla_n x$$

a bounded operator on X.

For C_0-semigroups which are not analytic we have the following version of the Hille–Yosida theorem (see [127]).

Theorem 2.5.9. *A C_0-semigroup $T(t)$ with generator A on a Banach space X is R-bounded with $R(\{T(t) : t > 0\}) \leq C$ if and only if the set $T = \{\lambda^n R(\lambda, A)^n : \lambda > 0\}$ satisfies $R(T) \leq C$.*

Proposition 2.5.10 ([22]). *Let $T \in \mathcal{B}(X)$ be power bounded and analytic. Then $\{T^n, n(T - I)T^n : n \in \mathbb{N}\}$ is R-bounded if $\{(\lambda - 1)R(\lambda, T) : |\lambda| = 1, \lambda \neq 1\}$ is R-bounded.*

The converse of Blunck's theorem holds without restriction on the Banach space X, in the following sense.

Proposition 2.5.11. *Let* $p \in (1, \infty)$ *and let* X *be a Banach space. Let* $M :$ $(-\pi, \pi) \setminus \{0\} \to \mathcal{B}(X)$ *be a function. Suppose that there is an operator* $T_M \in$ $\mathcal{B}(l_p(\mathbb{Z}; X))$ *such that*

$$\mathcal{F}(T_M f)(e^{it}) = M(t)\mathcal{F}f(e^{it}), \quad t \in (-\pi, \pi) \setminus \{0\}, \ \mathcal{F}f \in L^\infty(\mathbb{T}; X)$$

of compact support. Then the set

$$\{M(t) : t \in (-\pi, \pi)\}$$

is R-bounded.

This result is a consequence of [22, Proposition 1.4].

Since its conception in the mid-1990s, R-boundedness has proved to be an important tool in the theory of maximal regularity of evolution equations, in operator theory, Schauder decompositions, vector-valued harmonic analysis, partial differential equations, pseudo differential operators, and stochastic equations (see [19, 45, 66, 67, 88, 113, 176, 177, 180]).

Maximal Regularity in Hardy Spaces

Hytönen in [101] has considered maximal regularity in Hardy spaces (see Chap. 7 for notations). He proved the following result.

Theorem 2.5.12. *Let* X *be a UMD space and let* A *be a closed, linear, densely defined operator on* X. *Then the following assertions are equivalent:*

(C_1) *Equation (2.3.1) has maximal* L^p-*regularity on* $[0, \infty)$ *for all* $p \in (1, \infty)$.
(C_2) *Equation (2.3.1) has maximal* H^1-*regularity on* $[0, \infty)$.
(C_3) *Equation (2.3.1) has* (H^1, L^1)-*regularity on* $[0, \infty)$.
(C_4) $-A$ *generates a bounded analytic semigroup and*

$$\{A(i2\pi\xi + A)^{-1} : \xi \in \mathbb{R}\setminus\{0\}\}$$

is R-bounded.
Moreover, any of these is sufficient to
(C_5) *Equation (2.3.1) has maximal* H^p-*regularity on* $[0, \infty)$ *for all* $p \in (0, 1)$.

Remark 2.5.13. The implications $(C_1) \Rightarrow (C_2) \Rightarrow (C_3) \Rightarrow (C_4)$ and $(C_2) \Rightarrow (C_5)$ hold in fact for any Banach space X.

Remark 2.5.14. By (H^1, L^1)-regularity of (2.3.1) on $[0, \infty)$ we mean that for every $f \in H^1([0, \infty); X)$ there exists a unique $u \in W_{loc}^{1,1}([0, \infty); X)$ such that $u(t) \in$ $D(A)$ and $u'(t) = Au(t) + f(t)$ for a.e. $t > 0$ and moreover

$$\|u'\|_{L^1([0,\infty);X)} + \|Au\|_{L^1([0,\infty);X)} \le C\|f\|_{H^1([0,\infty);X)},$$

where $C < \infty$ is independent of f. For the definition of maximal H^1-regularity, replace L^1 by H^1.

Concerning maximal H^p-regularity when $0 < p < 1$, we simply require that the map $f \to Au$ is well defined and bounded on a dense subspace of $H^p([0,\infty); X)$ consisting of proper functions.

Multiplier Theorems

We have the following special case of the Marcinkiewicz multiplier theorem (Theorem 2.4.8).

Theorem 2.5.15. Let $X = L^{p_1}(\Omega, \Sigma, \mu)$, $Y = L^{p_2}(\Omega, \Sigma, \mu)$ where $1 < p_1 \le 2 \le p_2 < \infty$ and (Ω, Σ, μ) is a measure space. Then each bounded sequence $\{M_k\}_{k \in \mathbb{Z}} \subset \mathcal{B}(X, Y)$ satisfying

$$\sup_{k \in \mathbb{Z}} \|k(M_{k+1} - M_k)\| < \infty$$

is an L^p-multiplier for $1 \le p < \infty$ (see [8]).

The following proposition shows that we cannot replace the R-boundedness in Theorem 2.4.8 by boundedness in operator norm unless the underlying Banach spaces X is of cotype 2 and Y is of type 2. When $X = Y$, this is equivalent to saying that X is isomorphic to a Hilbert space (see [8] for details).

Proposition 2.5.16. Let X and Y be UMD spaces. Then the following assertions are equivalent:

(a) X is of cotype 2 and Y is of type 2.
(b) There is $1 < p < \infty$ such that each sequence $\{M_k\}_{k \in \mathbb{Z}} \subset \mathcal{B}(X, Y)$ satisfying $\sup_{k \in \mathbb{Z}} \|M_k\| < \infty$ and $\sup_{k \in \mathbb{Z}} \|k(M_{k+1} - M_k)\| < \infty$ is an L^p-multiplier.

Remark 2.5.17. In the scalar case more general conditions are known to be sufficient in Theorem 2.4.8. Let $\{M_k\}_{k \in \mathbb{Z}}$ be a bounded scalar sequence. Instead of assuming that $k(M_{k+1} - M_k)$ is bounded, it suffices to assume that

$$\sup_{j \in \mathbb{N}} \sum_{2^j \le |k| < 2^{j+1}} |M_{k+1} - M_k| < \infty,$$

in order to deduce that $\{M_k\}_{k \in \mathbb{Z}}$ is an L^p-multiplier (see[8, 74]).

Definition 2.5.18. (Dyadic decomposition). (a) We decompose $(0, \pi)$ into the following family $(I_j)_{j \in \mathbb{Z}}$ of intervals:

$$I_j := [\pi - 2^{-(j+1)}\pi, \pi - 2^{-(j+2)}\pi), \ j \geq 0,$$

$$I_j := [2^{j-1}\pi, 2^j\pi), \ j < 0.$$

(b) Now we denote by a_j, b_j, Δ_j the endpoints and the corresponding arcs of I_j:

$$I_j = [a_j, b_j), \text{ and } \Delta_j := \{e^{it} : t \in -I_j \cup I_j\}.$$

We have the following Marcinkiewicz-type multiplier theorem.

Theorem 2.5.19 ([22]). *Let X be a UMD space and $1 < p < \infty$. Then, for all $M : \mathbb{T} \to \mathcal{B}(X)$ of the form $M = \sum_{j \in \mathbb{Z}} \chi_{\Delta_j} m M_j$, where $m : \mathbb{T} \to \mathbb{C}$ has uniformly bounded variations over $(\Delta_j)_{j \in \mathbb{Z}}$ and $\{M_j : j \in \mathbb{Z}\} \subset \mathcal{B}(X)$ is R-bounded, we have*

$$\|T_M\|_{\mathcal{B}(l^p(\mathbb{Z}, X))} \leq C_{p,X} R(\{M_j : j \in \mathbb{Z}\}) \sup_{j \in \mathbb{Z}} Var_{\Delta_j} m.$$

Here we write $Var_{\Delta_j} m := \max\{\text{var}_{\Delta_j} m, \|m\|_{L^\infty(\Delta_j)}\}$, where $\text{var}_{\Delta_j} m$ is the usual variation of m over Δ_j.

We remark that Witvliet's thesis [181] contains an extensive treatment of multiplier theorems and its applications. In [174] the classical Fourier multiplier theorems of Marcinkiewicz and Miklin are extended to vector-valued functions and operator-valued multiplier functions on $\mathbb{Z}^d$ or $\mathbb{R}^d$ which satisfy certain R-boundedness conditions.

We say that a function $M : \mathbb{R}^d \setminus \{0\} \to \mathcal{B}(X, Y)$ is a Fourier multiplier, i.e., $M \in \mathcal{M}_p(\mathbb{R}^d; X, Y)$ if the operator (2.4.3) first defined for $f \in \mathcal{S}(\mathbb{R}^d; X)$ extends to a bounded operator from $L^p(\mathbb{R}^d; X)$ to $L^p(\mathbb{R}^d; Y)$.

We have the following two Miklin-type Fourier theorems due to Štrkalj and Weis [174].

Theorem 2.5.20. *Let X and Y be UMD spaces and $1 < p < \infty$. If the function $M : \mathbb{R}^d \setminus \{0\} \to \mathcal{B}(X, Y)$ has the property that their distributional derivatives $D^\gamma M$ of the order $\gamma \leq (1, \cdots, 1)$ are represented by functions and moreover*

$$R\left(\left\{|x|^{|\gamma|}(D^\gamma M)(x) : x \in \mathbb{R}^d \setminus \{0\}, \gamma \leq (1, \ldots, 1)\right\}\right) < \infty,$$

holds, then $M \in \mathcal{M}_p(\mathbb{R}^d; X, Y)$.

A Banach space X has the property (α) if there is a constant C_X such that for each $n \in \mathbb{N}$, subset $\{x_{i,j}\}_{i,j=1}^n$ of X, and subset $\{\alpha_{i,j}\}_{i,j=1}^n$ of $\mathbb{C}$ with $|\alpha_{i,j}| \leq 1$ we have

$$\int_0^1 \int_0^1 \left\| \sum_{i,j=1}^n r_i(u) r_j(v) \alpha_{i,j} x_{i,j} \right\|_X du\, dv$$

$$\leq C_X \int_0^1 \int_0^1 \left\| \sum_{i,j=1}^n r_i(u) r_j(v) x_{i,j} \right\|_X du\, dv,$$

where r_j are the Rademacher functions (see [127]).

Theorem 2.5.21. *Let X and Y be UMD spaces with the property (α) and where the function $M : \mathbb{R}^d \setminus \{0\} \to \mathcal{B}(X, Y)$ has the property that their distributional derivatives $D^\gamma M$ of the order $\gamma \leq (1, \cdots, 1)$ are represented by functions which fulfill*

$$R\left(\left\{ x^\gamma (D^\gamma M)(x) : x \in \mathbb{R}^d \setminus \{0\}, \gamma \leq (1, \ldots, 1) \right\}\right) < \infty,$$

then $M \in \mathcal{M}_p(\mathbb{R}^d; X, Y)$ $(1 < p < \infty)$.

Fourier Multiplier on Besov Spaces

Besov spaces are first introduced and investigated by Besov [20]. These spaces occur naturally in many fields of analysis. They can be defined via dyadic decomposition and form scales $B^s_{p,q}$ carrying three indices $s \in \mathbb{R}$, $1 \leq p, q \leq \infty$. We briefly recall the definition and some of the properties of periodic Besov spaces in the vector-valued case introduced in [9]. For the scalar case see [168, 175].

Let $S(\mathbb{R})$ be the Schwartz space of all rapidly decreasing smooth functions on $\mathbb{R}$. In order to define Besov spaces, we consider the dyadic-like subsets of $\mathbb{R} : I_0 := \{t \in \mathbb{R} : |t| \leq 2\}$, $I_k := \{t \in \mathbb{R} : 2^{k-1} < |t| \leq 2^{k+1}\}$ for $k \in \mathbb{N}$. Let $\Phi(\mathbb{R})$ be the set of all systems $\phi = (\phi_k)_{k \in \mathbb{Z}_+} \subset S(\mathbb{R})$ satisfying $supp(\phi_k) \subset \bar{I}_k$ for each $k \in \mathbb{Z}_+$, $\sum_{k \in \mathbb{Z}_+} \phi_k(x) = 1$ for $x \in \mathbb{R}$, and for each $m \in \mathbb{Z}_+$,

$$\sup_{k \in \mathbb{Z}_+, x \in \mathbb{R}} 2^{km} |\phi_k^{(m)}(x)| < \infty.$$

The set $\Phi(\mathbb{R})$ is not empty [175]. For $f \in \mathcal{D}'([0, 2\pi]; X)$,[8] denote by $\hat{f}(k)$, for $k \in \mathbb{Z}$, the kth Fourier coefficient of f as $< \hat{f}, l > = < f, \hat{l} >$, $l \in \mathcal{D}([0, 2\pi])$. In what follows we identify $\hat{l}$ with $\hat{f}$ which is standard in the theory of Besov and Triebel spaces.

---

[8] $\mathcal{D}'([0, 2\pi]; X)$ is the set of all linear mappings T from $\mathcal{D}([0, 2\pi])$ into X such that $\|T(f)\|_X \leq C \sum_{n \leq N} \sup_{t \in [0, 2\pi]} |f^{(n)}(t)|$ for all $f \in \mathcal{D}([0, 2\pi])$ and for some $N \in \mathbb{N}$ and $C > 0$ independent of f. Elements in $\mathcal{D}'([0, 2\pi]; X)$ are called X-valued distributions on $[0, 2\pi]$. We use the weak topology on $\mathcal{D}'([0, 2\pi]; X)$, i.e., a sequence T_k converges to T in $\mathcal{D}'([0, 2\pi]; X)$ if and only if $\lim_{k \to \infty} T_k(f) = T(f)$ for all $f \in \mathcal{D}([0, 2\pi])$.

Let $1 \le p, q \le \infty$, $s \in \mathbb{R}$ and let $\phi = (\phi_k)_{k \in \mathbb{Z}_+} \in \Phi(\mathbb{R})$ be fixed. The X-valued periodic Besov space $B^s_{p,q}([0, 2\pi]; X)$ is defined by the set

$$\Big\{ f \in \mathcal{D}'([0, 2\pi]; X) : \|f\|_{B^s_{p,q}} :=$$

$$\Big(\sum_{j \ge 0} 2^{sjq} \Big\| \sum_{k \in \mathbb{Z}} e_k \otimes \phi_j(k) \hat{f}(k) \Big\|_p^q \Big)^{1/q} < \infty \Big\},$$

with the usual modification if $q = \infty$.

Remark 2.5.22 ([9]). (a) We note that the space $B^s_{\infty, \infty}$ is the familiar space of all Hölder continuous functions of index s if $s \in (0, 1)$.

(b) The space $B^s_{p,q}([0, 2\pi]; X)$ is independent from the choice of ϕ and different choices of ϕ lead to equivalent norms $\| \cdot \|_{B^s_{p,q}}$ on $B^s_{p,q}([0, 2\pi]; X)$.

(c) Note that for $f \in \mathcal{D}'([0, 2\pi]; X)$ and $j \in \mathbb{Z}$, $\sum_{k \in \mathbb{Z}} e_k \otimes \phi_j(k) \hat{f}(k)$ is a trigonometric polynomial. Then

$$\Big\| \sum_{k \in \mathbb{Z}} e_k \otimes \phi_j(k) \hat{f}(k) \Big\|_p < \infty, \quad 1 \le p \le \infty.$$

We can identify $B^s_{p,q}([0, 2\pi]; X)$ with the space of all sequences $(a_k)_{k \in \mathbb{Z}}$ in X such that

$$\Big(2^{sjq} \Big\| \sum_{k \in \mathbb{Z}} e_k \otimes \phi_j(k) a_k \Big\|_p \Big)_{j \ge 0} \in l^q.$$

This shows that $\| \cdot \|_{B^s_{p,q}}$ is a norm and that $B^s_{p,q}([0, 2\pi]; X)$ equipped with the norm $\| \cdot \|_{B^s_{p,q}}$ is a Banach space.

(d) For $p, q < \infty$, the set of all X-valued trigonometric polynomials is dense in $B^s_{p,q}([0, 2\pi]; X)$.

Next, we summarize some useful properties of Besov spaces [9].

Remark 2.5.23. (a) The natural injection from $B^s_{p,q}([0, 2\pi]; X)$ into $L^p(0, 2\pi; X)$ is a continuous linear operator for $s > 0$.

(b) The natural injection from $B^{s+\varepsilon}_{p,q}([0, 2\pi]; X)$ into $B^s_{p,q}([0, 2\pi]; X)$ is a continuous linear operator for $\varepsilon > 0$.

(c) Lifting property: let $f \in \mathcal{D}'([0, 2\pi]; X)$ and $\eta \in \mathbb{R}$. Then $f \in B^s_{p,q}([0, 2\pi]; X)$ if and only if

$$\sum_{k \neq 0} e_k \otimes k^\eta \hat{f}(k) \in B_{p,q}^{s-\eta}([0, 2\pi]; X).$$

(d) Let $s > 0$. Then $f \in B_{p,q}^{1+s}([0, 2\pi]; X)$ if and only if f is differentiable a.e. and $f' \in B_{p,q}^{s}([0, 2\pi]; X)$.

Remark 2.5.24 ([9]). Many applications of Besov spaces require a knowledge of their interpolation properties: Let $1 \leq p, q_0, q_1, q \leq \infty$, $s_0, s_1 \in \mathbb{R}$, $s_0 \neq s_1$, $0 < \theta < 1$ and $s = \theta s_1 + (1 - \theta)s_0$. Then

$$(B_{p,q_0}^{s_0}([0, 2\pi]; X), B_{p,q_1}^{s_1}([0, 2\pi]; X))_{\theta,q} = B_{p,q}^{s}([0, 2\pi]; X).$$

The following definition was introduced in [120].

Definition 2.5.25. A sequence $\{M_k\}_{k \in \mathbb{Z}} \subset \mathcal{B}(X, Y)$ satisfies a Marcinkiewicz estimate of order 1 if

$$\sup_{k \in \mathbb{Z}} ||M_k|| < \infty, \quad \sup_{k \in \mathbb{Z}} ||k(M_{k+1} - M_k)|| < \infty. \qquad (2.5.1)$$

If in addition we have

$$\sup_{k \in \mathbb{Z}} ||k^2(M_{k+1} - 2M_k + M_{k-1})|| < \infty, \qquad (2.5.2)$$

then we say that $\{M_k\}_{k \in \mathbb{Z}}$ satisfies a Marcinkiewicz estimate of order 2. Finally, if in addition to (2.5.1) and (2.5.2) we have

$$\sup_{k \in \mathbb{Z}} ||k^3(M_{k+1} - 3M_k + 3M_{k-1} - M_{k-2})|| < \infty, \qquad (2.5.3)$$

then we say that $\{M_k\}_{k \in \mathbb{Z}}$ satisfies a Marcinkiewicz estimate of order 3.

Remark 2.5.26. (a) If $\{M_k\}_{k \in \mathbb{Z}}$ and $\{N_k\}_{k \in \mathbb{Z}}$ satisfy Marcinkiewicz estimate of order k ($k = 1, 2, 3$), then $\{M_k \pm N_k\}_{k \in \mathbb{Z}}$ satisfy Marcinkiewicz estimate of the same order.

(b) (Scalar case) If $\{a_k\}_{k \in \mathbb{Z}}$ and $\{b_k\}_{k \in \mathbb{Z}}$ are sequences that satisfy Marcinkiewicz estimate of order k ($k = 1, 2, 3$), then $\{a_k b_k\}_{k \in \mathbb{Z}}$ satisfy Marcinkiewicz estimate of the same order.

The proof of the above properties as well as other related consequences can be found in [120].

In the study of maximal regularity on periodic Besov spaces for first (resp. second)-order integrodifferential equations with infinite delay and for second-order delay equations the use of operator-valued $B_{p,q}^{s}$-multiplier technique plays an important role (see [27, 117, 159]).

Definition 2.5.27. Let X and Y be two Banach spaces, $1 \le p, q \le \infty$, $s \in \mathbb{R}$ and let $\{M_k\}_{k \in \mathbb{Z}}$ be a sequence in $\mathcal{B}(X, Y)$. We say that $\{M_k\}_{k \in \mathbb{Z}}$ is a $B_{p,q}^s$-multiplier if for each $f \in B_{p,q}^s([0, 2\pi]; X)$ there is $u \in B_{p,q}^s([0, 2\pi]; Y)$, such that $\hat{u}(k) = M_k \hat{f}(k)$ for all $k \in \mathbb{Z}$.

In this case, it follows from the closed graph theorem that there is $C > 0$ such that for $f \in B_{p,q}^s([0, 2\pi]; X)$, we have

$$\left\| \sum_{k \in \mathbb{Z}} e_k \otimes M_k \hat{f}(k) \right\|_{B_{p,q}^s} \le C \|f\|_{B_{p,q}^s}.$$

There is a unique operator $M \in \mathcal{B}(B_{p,q}^s([0, 2\pi]; X), B_{p,q}^s([0, 2\pi]; Y))$ such that

$$\widehat{Mf}(k) = M_k \hat{f}(k), \quad \text{for all } k \in \mathbb{Z},$$

for all $f \in B_{p,q}^s([0, 2\pi]; X)$. We call M the operator associated with $\{M_k\}_{k \in \mathbb{Z}}$. One has

$$Mf = \lim_{n \to \infty} \frac{1}{n+1} \sum_{m=0}^{n} \sum_{k=-m}^{m} e_k \otimes M_k \hat{f}(k)$$

in $B_{p,q}^s([0, 2\pi]; Y)$ for all $f \in B_{p,q}^s([0, 2\pi]; X)$.

Remark 2.5.28. Let X, Y, and Z be Banach spaces. If $\{M_k\}_{k \in \mathbb{Z}} \subset \mathcal{B}(X, Y)$ and $\{N_k\}_{k \in \mathbb{Z}} \subset \mathcal{B}(Y, Z)$ are $B_{p,q}^s$-multiplier, then $\{N_k M_k\}_{k \in \mathbb{Z}}$ is a $B_{p,q}^s$-multiplier. This follows directly from the definition.

The following result gives a sufficient condition for an operator-valued sequence to be a $B_{p,q}^s$-multiplier (see [9]).

Theorem 2.5.29. *Let X and Y be Banach spaces and let $\{M_k\}_{k \in \mathbb{Z}}$ be in $\mathcal{B}(X, Y)$ such that it satisfies Marcinkiewicz estimate of order 2. Then for $1 \le p, q \le \infty$, $s \in \mathbb{R}$, $\{M_k\}_{k \in \mathbb{Z}}$ is a $B_{p,q}^s$-multiplier. Moreover, if X and Y are B-convex,[9] then the Marcinkiewicz estimate of order 1 is sufficient for $\{M_k\}_{k \in \mathbb{Z}}$ to be a $B_{p,q}^s$-multiplier.*

Remark 2.5.30. Let X be an arbitrary Banach space and let $1 \le p, q \le \infty$, $s \in \mathbb{R}$. Let $M_k = I$ for $k \ge 0$ and $M_k = 0$ for $k < 0$. Then $\{M_k\}_{k \in \mathbb{Z}}$ defines a $B_{p,q}^s$-multiplier by Theorem 2.5.29. The associated operator is called the Riesz projection. Similarly, letting $M_k = i(sign(k))I$ defines a $B_{p,q}^s$-multiplier. The associated operator is the Hilbert transform.

---

[9] We recall that a Banach space X is B-convex if it does not contain l_1^n uniformly. This is equivalent to saying that X has Fourier type $1 < p \le 2$, i.e., the Fourier transform is a bounded linear operator from $L^p(0, 2\pi; X)$ into $l^q(\mathbb{Z}, X)$ where $1/p + 1/q = 1$.

Fourier Multiplier on Triebel–Lizorkin Spaces

In the Triebel–Lizorkin space case, the results concerning Fourier multipliers are similar to that on Besov spaces obtained by Arendt and Bu [9], i.e., the results will not depend on the geometry of the underlying Banach space. However, one requires more on the smoothness of the multipliers (see Theorem 2.5.33). We recall the definition of periodic Triebel–Lizorkin spaces in the vector-valued case [28]. We use the same notations $\mathcal{S}(\mathbb{R})$, $\mathcal{D}([0, 2\pi])$, $\mathcal{D}'([0, 2\pi]; X)$, $\Phi(\mathbb{R})$ as in the preceding section. Let $\phi = (\phi_k)_{k \in \mathbb{Z}_+} \in \Phi(\mathbb{R})$ be fixed. For $1 \leq p < \infty$, $1 \leq q \leq \infty$, $s \in \mathbb{R}$, the X-valued periodic Triebel–Lizorkin space $F_{p,q}^s([0, 2\pi]; X)$ is defined by

$$\left\{ f \in \mathcal{D}'([0, 2\pi]; X) : \|f\|_{F_{p,q}^s} := \left\| \left(\sum_{j \geq 0} 2^{sjq} \left\| \sum_{k \in \mathbb{Z}} e_k \otimes \phi_j(k) \hat{f}(k) \right\|_X^q \right)^{1/q} \right\|_p < \infty \right\},$$

with the usual modification if $q = \infty$.

The space $F_{p,q}^s([0, 2\pi]; X)$ is independent from the choice of ϕ and different choices of ϕ lead to equivalent norm $\| \cdot \|_{F_{p,q}^s}$ on $F_{p,q}^s([0, 2\pi]; X)$. The Triebel–Lizorkin space $F_{p,q}^s([0, 2\pi]; X)$ equipped with the norm $\| \cdot \|_{F_{p,q}^s}$ is a Banach space.

The following proposition summarizes the elementary properties of Triebel–Lizorkin spaces. Here "$\hookrightarrow$" means that the natural inclusion is continuous (see [28, Proposition 2.3]).

Proposition 2.5.31. *The following properties hold:*

(a) $\mathcal{D}([0, 2\pi]; X) \hookrightarrow F_{p,q}^s([0, 2\pi]; X) \hookrightarrow \mathcal{D}'([0, 2\pi]; X)$.
(b) *For $q < \infty$, the set of all X-valued trigonometric polynomials is dense in $F_{p,q}^s$.*
(c) *If $s > 0$, then $F_{p,q}^s \hookrightarrow L^p$.*
(d) *If $1 \leq p_0 \leq p_1 < \infty$, $1 \leq q \leq \infty$, $s \in \mathbb{R}$, then $F_{p_1,q}^s \hookrightarrow F_{p_0,q}^s$.*
(e) *If $1 \leq q_0 \leq q_1 \leq \infty$, $1 \leq p < \infty$, $s \in \mathbb{R}$, then $F_{p,q_0}^s \hookrightarrow F_{p,q_1}^s$.*
(f) *If $1 \leq q_0, q_1 \leq \infty$, $1 \leq p < \infty$, $s \in \mathbb{R}$ and $\varepsilon > 0$, then $F_{p,q_0}^{s+\varepsilon} \hookrightarrow F_{p,q_1}^s$.*
(g) *Let $f \in \mathcal{D}'([0, 2\pi]; X)$ and $\eta \in \mathbb{R}$. Then $f \in F_{p,q}^s$ if and only if*

$$\sum_{k \neq 0} e_k \otimes k^\eta \hat{f}(k) \in F_{p,q}^{s-\eta}.$$

(h) *Let $s > 0$. Then $f \in F_{p,q}^{1+s}$ if and only if f is differentiable a.e. and $f' \in F_{p,q}^s$.*

Definition 2.5.32. Let X and Y be two Banach spaces, $1 \leq p < \infty$, $1 \leq q \leq \infty$, $s \in \mathbb{R}$ and let $\{M_k\}_{k \in \mathbb{Z}}$ be a sequence in $\mathcal{B}(X, Y)$. We say that $\{M_k\}_{k \in \mathbb{Z}}$ is an $F_{p,q}^s$-multiplier if for each $f \in F_{p,q}^s([0, 2\pi]; X)$ there is $u \in F_{p,q}^s([0, 2\pi]; Y)$, such that $\hat{u}(k) = M_k \hat{f}(k)$ for all $k \in \mathbb{Z}$.

In this case, it follows from the closed graph theorem that there is $C > 0$ such that for $f \in F_{p,q}^s([0, 2\pi]; X)$, we have

$$\left\|\sum_{k\in\mathbb{Z}} e_k \otimes M_k \hat{f}(k)\right\|_{F_{p,q}^s} \leq C \|f\|_{F_{p,q}^s}.$$

There is a unique operator $M \in \mathcal{B}(F_{p,q}^s([0,2\pi]; X), F_{p,q}^s([0,2\pi]; Y))$ associated with $\{M_k\}_{k\in\mathbb{Z}}$.

The following result gives a sufficient condition for an operator-valued sequence to be a $F_{p,q}^s$-multiplier (see [28]).

Theorem 2.5.33. *Let X and Y be Banach spaces, $1 \leq p < \infty$, $1 \leq q \leq \infty$, $s \in \mathbb{R}$ and let $\{M_k\}_{k\in\mathbb{Z}}$ be a sequence in $\mathcal{B}(X,Y)$ such that it satisfies Marcinkiewicz estimate of order 3. Then $\{M_k\}_{k\in\mathbb{Z}}$ is an $F_{p,q}^s$-multiplier. Moreover, if X and Y are B-convex, then the Marcinkiewicz estimate of order 2 is sufficient for $\{M_k\}_{k\in\mathbb{Z}}$ to be an $F_{p,q}^s$-multiplier.*

Remark 2.5.34. In [28] by using Theorem 2.5.33 the authors gave a characterization of the maximal regularity in the sense of Triebel spaces for Cauchy problems with periodic boundary conditions.

Remark 2.5.35. Let X be an arbitrary Banach space and let $1 \leq p < \infty$, $1 \leq q \leq \infty$, $s \in \mathbb{R}$. Let $M_k = I$ for $k \geq 0$ and $M_k = 0$ for $k < 0$. Then $\{M_k\}_{k\in\mathbb{Z}}$ defines a $F_{p,q}^s$-multiplier by Theorem 2.5.33. Similarly, letting $M_k = i(sign(k))I$ defines a $F_{p,q}^s$-multiplier.

Chapter 3
First-Order Linear Difference Equations

In this chapter we present the maximal discrete regularity approach to first-order linear difference equations in general Banach spaces. In the first section we introduce the general frame for first-order linear difference equations. The entire linear theory of maximal regularity is not only important on its own, but it is also the indispensable basis for the theory of nonlinear difference equations, which we present in the next chapter.

3.1 A Characterization of Maximal l_p-Regularity

Let $T \in \mathcal{B}(X)$ be given, define $\mathcal{T} : \mathbb{Z}_+ \to \mathcal{B}(X)$ by $\mathcal{T}(n) = T^n$ and consider the discrete-time evolution equation

$$\begin{cases} \Delta x_n - (T - I)x_n = f_n, n \in \mathbb{Z}_+, \\ \\ x_0 = 0, \end{cases} \tag{3.1.1}$$

The following definition was introduced by Blunck [23, p. 212].

Definition 3.1.1. Let $1 < p < +\infty$. Let $T \in \mathcal{B}(X)$ be a power-bounded operator. We say that (3.1.1) has discrete maximal l_p-regularity if

$$(\mathcal{K}_T f)(n) := \sum_{k=0}^{n}(T - I)\mathcal{T}(k)f_{n-k}$$

defines a bounded operator $\mathcal{K}_T \in \mathcal{B}(l_p(\mathbb{Z}_+; X))$ for some $p \in (1, \infty)$.

In other words, the question is if $f \in l_p(\mathbb{Z}_+, X)$ implies $\Delta x_n \in l_p(\mathbb{Z}_+, X)$ (cf. Chap. 1, Sect. 1.3).

R.P. Agarwal et al., *Regularity of Difference Equations on Banach Spaces*,
DOI 10.1007/978-3-319-06447-5_3, © Springer International Publishing Switzerland 2014

Remark 3.1.2. Just as in the continuous case, discrete maximal l_p-regularity for $1 < p < \infty$ turns out to be independent on p [22].

Remark 3.1.3. It was shown by Blunck [22] that a necessary condition for (3.1.1) has discrete maximal l_p-regularity for some p is that T satisfies Ritt's condition (1.5.1), whence T is analytic (see Definition 1.3.7).[1]

In [23], Blunck characterizes the discrete maximal l_p-regularity for first-order difference equations by R-boundedness properties of the resolvent operator T as follows.

Theorem 3.1.4. *Let X be a UMD space and let $T \in \mathcal{B}(X)$ be power bounded and analytic. Then the following assertions are equivalent.*

(i) *Equation (3.1.1) has discrete maximal l_p-regularity.*

(ii) *The set $\left\{ (z-1)\left(z-T\right)^{-1} : |z| = 1, \, z \neq 1 \right\}$ is R-bounded.*

Proof. Since T is power bounded and analytic, we have by Theorem 1.3.9

$$\sigma(T) \subset \{z \in \mathbb{C} : |z| < 1\} \cup \{1\}.$$

In particular, $(z - T)^{-1}$ exists for $|z| = 1, z \neq 1$. (ii) $\implies$ (i). By definition, (3.1.1) has discrete maximal l_p-regularity if and only if the operator $\mathcal{K}_T f(n) := \sum_{k=0}^{n}(T - I)T(k) f_{n-k}$ is bounded on $l_p(\mathbb{Z}_+; X)$ for some $p \in (1, \infty)$. Define $\mathcal{K}_T f(n)$ by 0 for $n < 0$. Taking the Z-transform to $\mathcal{K}_T f$, we obtain

$$\widehat{\mathcal{K}_T f}(z) = (T - I)z(z - T)^{-1} \hat{f}(z).$$

From the identity $T(z - T)^{-1} = z(z - T)^{-1} - I$ we obtain

---

[1]If (3.1.1) has discrete maximal l_p-regularity, then T is analytic. In fact, put $M := \sup_n \|T^n\|$, we consider for all $b \in \mathbb{N}$ and $x \in X$ the sequence $f \in l_p(\mathbb{Z}_+, X)$ defined by $f_j = T^j x$ for $j = 1, \ldots, b$ and $f_j = 0$ otherwise. We have the following estimates:

$$\|(T - I)T^b x\|_X \leq \|T^{b-n}(T - I)T^n x\|_X \leq \|T^{b-n}\|\|(T - I)T^n x\|_X \leq M\|(T - I)T^n x\|_X,$$

$$\|\mathcal{K}_T f\|_p \leq \|\mathcal{K}_T\|_{\mathcal{B}(l_p(\mathbb{Z}_+;X))}\|f\|_p \leq \|\mathcal{K}_T\|_{\mathcal{B}(l_p(\mathbb{Z}_+;X))} M b^{1/p} \|x\|_X.$$

Using the first one, we get

$$\|\mathcal{K}_T f\|_p \geq \left(\sum_{n=1}^{b} \|(\mathcal{K}_T f)_n\|_X^p \right)^{1/p} = \left(\sum_{n=1}^{b} n^p \|(T - I)T^n x\|_X^p \right)^{1/p}$$

$$\geq M^{-1} \left(\sum_{n=1}^{b} n^p \right)^{1/p} \|(T - I)T^b x\|_X \geq (2M)^{-1} b^{1+1/p} \|(T - I)T^b x\|_X.$$

Therefore $\|b(T - I)T^b x\|_X \leq 2M^2 \|\mathcal{K}_T\|_{\mathcal{B}(l_p(\mathbb{Z}_+;X))} \|x\|_X.$

$$(T - I)z(z - T)^{-1} = zT(z - T)^{-1} - z(z - T)^{-1}$$

$$= z\big(z(z - T)^{-1} - I\big) - z(z - T)^{-1}$$

$$= z\big(z(z - T)^{-1} - (z - T)^{-1} - I\big)$$

$$= z\big((z - 1)(z - T)^{-1} - I\big).$$

Hence

$$\widehat{\mathcal{K}_T f}(z) = z\big((z - 1)(z - T)^{-1} - I\big)\hat{f}(z). \qquad (3.1.2)$$

By hypothesis, the set

$$\left\{(z - 1)\big(z - T\big)^{-1} : |z| = 1, z \neq 1\right\}$$

is R-bounded. Define

$$M(t) := (e^{it} - 1)(e^{it} - T)^{-1}, t \in (0, 2\pi).$$

Then

$$(e^{it} - 1)(e^{it} + 1)M'(t) = i e^{it}(e^{it} + 1)M(t) - i e^{it}(e^{it} + 1)(M(t))^2, \quad t \in (0, 2\pi).$$

We conclude by hypothesis and the permanence properties of R-bounded sets (Proposition 2.2.5) that the sets

$$\left\{M(t) : t \in (0, 2\pi)\right\} \text{ and } \left\{(e^{it} - 1)(e^{it} + 1)M'(t) : t \in (0, 2\pi)\right\}$$

are R-bounded.

Then, by Blunck's theorem (Theorem 2.4.9), there is an operator $T_M \in \mathcal{B}(l_p(\mathbb{Z}; X))$ such that

$$\widehat{T_M f}(z) = (z - 1)(z - T)^{-1}\hat{f}(z).$$

Hence (3.1.2) implies that

$$\widehat{\mathcal{K}_T f}(z) = z\widehat{T_M f}(z) - z\hat{f}(z).$$

Note that, explicitly $\mathcal{K}_T f(n) = T_M f(n + 1) - f(n + 1)$. Then by uniqueness of the Z-transform, $\mathcal{K}_T$ is bounded on $l_p(\mathbb{Z}_+; X)$, proving (i).

(i) $\Longrightarrow$ (ii). Define

$$Kf(n) := (\mathcal{K}_T f)(n-1) + f(n), \ f \in l_p(\mathbb{Z}; X), \ n \in \mathbb{Z}$$

and $\mathcal{K}_T f \mid \mathbb{Z}^- \equiv 0$. Then K is a bounded operator in $l_p(\mathbb{Z}; X)$. Indeed $\|K\|_{l_p(\mathbb{Z};X)} \le \|\mathcal{K}_T\|_{l_p(\mathbb{Z}_+;X)} + 1$. Next, define $\mathcal{T} : \mathbb{Z} \to \mathcal{B}(X)$ by $\mathcal{T}(n) = T^n$, $n \in \mathbb{Z}_+$, and $\mathcal{T} \mid \mathbb{Z}^- \equiv 0$. We observe that

$$\mathcal{F}[\mathcal{K}_T f(\cdot - 1)](z) = z^{-1}\mathcal{F}[\mathcal{K}_T f](z)$$

$$= z^{-1}\mathcal{F}[(T - I)(\mathcal{T} * f)](z)$$

$$= z^{-1}(T - I)\mathcal{F}[\mathcal{T} * f](z)$$

$$= z^{-1}(T - I)\mathcal{F}[\mathcal{T}](z)\mathcal{F}[f](z) \qquad (3.1.3)$$

$$= z^{-1}(T - I)Z[\mathcal{T}](z)\mathcal{F}[f](z)$$

$$= z^{-1}(T - I)z(z - T)^{-1}\mathcal{F}[f](z)$$

$$= (T - I)(z - T)^{-1}\mathcal{F}[f](z)$$

In view of (3.1.3) we have

$$\mathcal{F}[\mathcal{K} f](z) = \mathcal{F}[\mathcal{K}_T(\cdot - 1) f](z) + \mathcal{F}[f](z)$$

$$= (T - I)(z - T)^{-1}\mathcal{F}[f](z) + \mathcal{F}[f](z)$$

$$= ((T - I)(z - T)^{-1} + I)\mathcal{F}[f](z)$$

$$= (z^{-1}(T - I)z(z - T)^{-1} + I)\mathcal{F}[f](z)$$

$$= ((z - 1)(z - T)^{-1} - I + I)\mathcal{F}[f](z)$$

$$= (z - 1)(z - T)^{-1}\mathcal{F}[f](z).$$

It follows from Proposition 2.5.11 that the set

$$\left\{ (z - 1)\left(z - T\right)^{-1} : |z| = 1, z \ne 1 \right\}$$

is R-bounded. $\square$

3.2 Maximal l_p-Regularity for $p \in \{1, 2, \infty\}$

We begin this section with the following observation.

Remark 3.2.1. Let X be Banach space and let $(T^n)_{n \in \mathbb{Z}_+}$ be a discrete-time bounded analytic semigroup.

(a) The discrete Cauchy problem (3.1.1) has maximal l_2-regularity on X if and only if the discrete dual Cauchy problem

$$
\begin{cases}
\Delta x_n - (T^* - I)x_n = f_n, n \in \mathbb{Z}_+, \\
\\
x_0 = 0,
\end{cases}
\tag{3.2.1}
$$

has maximal l_2-regularity on X^*.

(b) The discrete Cauchy problem (3.1.1) has maximal l_1-regularity on X if and only if the discrete dual Cauchy problem (3.2.1) has maximal l_∞-regularity on X^*.

Definition 3.2.2. A Banach space X is said to have the discrete maximal regularity property $(DMRP)$ if for every discrete-time bounded analytic semigroup the associated discrete Cauchy problem (3.1.1) has maximal l_2-regularity.

Definition 3.2.3. A discrete-time analytic semigroup is called R-analytic (resp. R^*-analytic) if the set

$$
\{T^n, n(T^{n+1} - T^n) : n \in \mathbb{N}\}
$$

(resp. $\{T^{*n}, n(T^{*(n+1)} - T^{*n}) : n \in \mathbb{N}\}$) is R-bounded on X (resp. on X^*).

Definition 3.2.4. A Banach space X is said to have the (AR) property (resp. the $(AR)^*$ property) if for every discrete-time bounded analytic semigroup is R-analytic (resp. R^*-analytic).

Remark 3.2.5. The results of Blunck [22] show that the R-analyticity of a discrete-time bounded analytic semigroup is a necessary, and in UMD spaces sufficient, condition for (3.1.1) to have maximal l_2-regularity.

Portal in [160] has shown the following result.

Theorem 3.2.6. *Let X be a Banach space with $(DMRP)$. Then X has both properties (AR) and $(AR)^*$.*

The Hilbert spaces are the only spaces with $(DMRP)$ among spaces with an unconditional basis (see [160]). More precisely, we have the following theorem due to Kalton and Lancien [112].

Theorem 3.2.7. *Let X be a Banach space with $(DMRP)$ and an unconditional basis; then X is isomorphic to a Hilbert space.*

In the continuous-time setting (see Theorem 2.5.1) the existence of semigroup with unbounded generator such that the associated Cauchy problem has maximal L^∞-regularity is equivalent to the condition $c_0 \subset X$ in separable Banach spaces. In the discrete-time setting, Portal [160] has obtained the following results.

Theorem 3.2.8. *Let X be a separable Banach space. The following assertions are equivalent.*

(a) $c_0 \subset X$.
(b) *There is a bounded operator T acting on X and verifying the condition*

$$\exists\, C > 0 \text{ such that } ||n(T^{n+1} - T^n)|| \geq C \,\,\forall\, n \in \mathbb{Z}_+,$$

such that (3.1.1) has maximal l_∞-regularity.

Remark 3.2.9. The separability of X is only used to prove that (a) implies (b).

Corollary 3.2.10. *Let X be a Banach space such that $c_0 \not\subset X$ and let $(T^n)_{n \in \mathbb{Z}_+}$ be a discrete-time bounded analytic semigroup such that (3.1.1) has maximal l_∞-regularity. Then we have*

$$||n(T^{n+1} - T^n)|| \to 0, \text{ as } n \to \infty.$$

In the continuous setting, Guerre-Delabrière (Theorem 2.5.2) proved the following dual result of Theorem 2.5.1: $l_1 \subset_c X$ if and only if there is a bounded analytic semigroup with an unbounded generator such that (2.3.1) has maximal L^1-regularity. In a similar way in [160] the following result is proved.

Theorem 3.2.11. *Let X be a Banach space. The following assertions are equivalent.*

(a) $l_1 \subset_c X$.
(b) *There is a bounded operator T acting on X and verifying the condition*

$$\exists\, C > 0 \text{ such that } ||n(T^{n+1} - T^n)|| \geq C \,\,\forall\, n \in \mathbb{Z}_+,$$

such that (3.1.1) has maximal l_1-regularity.

Corollary 3.2.12. *Let X be a Banach space such that it contains no complemented copy of l_1 and let $(T^n)_{n \in \mathbb{Z}_+}$ be a discrete-time bounded analytic semigroup such that (3.1.1) has maximal l_1-regularity. Then we have*

$$||n(T^{n+1} - T^n)|| \to 0, \text{ as } n \to \infty.$$

Theorem 3.2.13 ([111]). *Let T be a power-bounded operator on a Banach space X. Then the following conditions are equivalent:*

(a) *Equation (3.1.1) has maximal l_1-regularity.*
(b) *There is a constant $C > 0$ such that*

$$\sum_{k=1}^{\infty} \left\| (T^k - T^{k-1})x \right\| \le C \|x\|, \ x \in X.$$

The corresponding result for maximal l_∞-regularity is the following.

Theorem 3.2.14 ([111]). *Let T be a power-bounded operator on a Banach space X. Then the following conditions are equivalent:*

(a) *Equation (3.1.1) has maximal l_∞-regularity.*
(b) *T satisfies Ritt's condition (1.5.1), and there is a constant $C > 0$ such that*

$$\|x\| \le C \left(\sup_{n \ge 1} \|n(T^n - T^{n-1})x\| + \limsup_{n \to \infty} \|T^n x\| \right).$$

The continuous analogue of the next theorem is well known [69, Theorem 7.1], which is due to Kalton and Portal [111].

Theorem 3.2.15. *Suppose (3.1.1) has either maximal l_1- or l_∞-regularity. Then (3.1.1) has maximal l_p-regularity for every $1 < p < \infty$.*

The next two results are the discrete analogues of the results of Baillon [14] and Guerre-Delabrière [90].

Theorem 3.2.16 ([111]). *Let $T \in B(X)$. Suppose that either*

(a) *X contains no copy of c_0 and (3.1.1) has maximal l_∞-regularity.*
(b) *X contains no complemented copy of l_1 and (3.1.1) has maximal l_1-regularity.*

Then X splits as a direct sum $X_1 \oplus X_2$ of T-invariant subspaces such that $T \mid_{X_1} = I_{X_1}$ and the spectral radius of $T \mid_{X_2}$ is strictly less than one.

Theorem 3.2.17 ([160]). *Let X be a Banach space and let $(T^n)_{n \in \mathbb{Z}_+}$ be a discrete-time bounded analytic semigroup such that*

(a) *T^n tends to zero in the strong operator topology.*
(b) *$1 \in \sigma(T)$.*
(c) *Equation (3.1.1) has maximal l_∞-regularity.*

Then $c_0 \subset X$.

We have the following dual corollary due to Portal [160].

Corollary 3.2.18. *Let X be a Banach space and let $(T^n)_{n \in \mathbb{Z}_+}$ be a discrete-time bounded analytic semigroup. Let us denote by S the adjoint operator of T acting on X^* and assume that:*

(a) *S^n tends to zero in the strong operator topology.*
(b) *$1 \in \sigma(T)$.*
(c) *Equation (3.1.1) has maximal l_1-regularity.*

Then $l_1 \subset_c X$.

3.3   Comments

Maximal regularity has also been studied in the finite difference setting by Portal [160–162], Ashyralyev et al. [12, 13], Geissert [86], Guidetti and Piskarev [91], Kalton and Portal [111], Castro et al. [37–39], and Cuevas et al. [52–55, 59]. In [160, 162], the author has discussed discrete analytic semigroup and maximal regularity on discrete-time scales, respectively. In [12], the authors have investigated well-posedness of difference schemes for abstract elliptic problems in $L_n^p([0, T], E_n)$ spaces under suitable R-boundedness condition when E_n is a UMD space. In [91] maximal regularity on discrete Hölder spaces for finite difference operators subject to Dirichlet boundary conditions in one and two dimensions is proved. Furthermore, the authors investigated maximal regularity in discrete Hölder spaces for the Crank–Nicolson scheme. In [86] maximal regularity for linear parabolic difference equations is treated, whereas in [53] a characterization in terms of R-boundedness properties of the resolvent operator for linear second-order difference equations was given (see Chap. 5). See also the recent paper by Kalton and Portal [111], where they discussed maximal regularity of power-bounded operators and relate the discrete to the continuous-time problem for analytic semigroups. Recently, discrete maximal regularity for functional difference equations with infinite delay was considered in [59] (see Chap. 7, Sect. 7.7).

Blunck in [22] also characterize the maximal regularity of (3.1.1) by the maximal regularity of the continuous-time evolution equation

$$\begin{cases} u'(t) = (T - I)u(t) + f(t), \ t > 0, \\ \\ u(0) = 0. \end{cases} \tag{3.3.1}$$

More precisely

Theorem 3.3.1. *Let X be a UMD space, and let $T \in \mathcal{B}(X)$ be power bounded and analytic. Then the following assertions are equivalent:*

  (i) *Equation (3.1.1) has discrete maximal l_p-regularity.*
 (ii) *$\{T^n, n(T - I)T^n : n \in \mathbb{N}\}$ is R-bounded.*
(iii) *Equation (3.3.1) has maximal regularity.*
 (iv) *$\{(\lambda - 1)R(\lambda, T) : \lambda \in 1 + i\mathbb{R}, \lambda \neq 1\}$ is R-bounded.*
  (v) *$\{e^{t(T-I)}, t(T - I)e^{t(T-I)} : t > 0\}$ is R-bounded.*

As an application of preceding result, Blunck [22] has obtained the following result.

Theorem 3.3.2. *Let $p \in (1, \infty)$ and let $T \in \mathcal{B}(L^p)$ be a subpositive analytic contraction.[2] Then (3.1.1) has discrete maximal l_p-regularity.*

---

[2]The subpositivity of a contraction T on L^p is defined by the existence of a dominating positive contraction S, i.e., $|Tf| \leq S|f|$ for all $f \in L^p$.

Remark 3.3.3. The preceding result shows that all Markov operators T have discrete maximal l_p-regularity on L^p for all $p \in (1, \infty)$ (see [22]).

Section 3.2 is taken from [111, 160, 161]. The proof of Remark 3.2.1 is contained in Portal's Thesis Lemma 4.2.7 and Lemma 4.4.2. In general if $1 \leq p \leq \infty$ and $\frac{1}{p} + \frac{1}{q} = 1$, then (3.1.1) has maximal l_p-regularity if and only if (3.2.1) has maximal l_q-regularity (see [111]).

Maximal regularity for nonautonomous difference equations is a topic that should be developed along the lines of the autonomous case (see Sect. 7.9). In the continuous setting, some advances in this direction are due to Amann [5], Saal, [166] and Hieber–Monniaux [98].

Chapter 4
First-Order Semilinear Difference Equations

The qualitative theory of difference equations is in the process of continuous development, as it is apparent from the huge number of research papers dedicated to it. Although several results in the discrete case are similar to those already known in the continuous case, the adaptation from the continuous to the discrete setting is not always direct and requires some special devices. Thus new challenges are faced.

In this section we prove the existence of bounded solutions whose first discrete derivative is in l_p, $(1 < p < \infty)$ for a first-order semilinear difference equation. For this, first we shall assume maximal regularity of the linear part to obtain a priori estimates; then we will use these estimates together with adequate assumptions on the nonlinearity to find a solution as a fixed point of a suitable operator defined in a discrete Sobolev space. The implementation of this approach is nontrivial, as the reader will perceive through this chapter.

4.1 Existence for the Semilinear Problem

In this section, our aim is to investigate the existence of bounded solutions, whose first discrete derivative is in ℓ_p, for semilinear difference equations via discrete maximal regularity.

Consider the following first-order difference equation:

$$\begin{cases} \Delta x_n - A x_n = f(n, x_n), n \in \mathbb{Z}_+, \\ \\ x_0 = 0, \end{cases} \tag{4.1.1}$$

R.P. Agarwal et al., *Regularity of Difference Equations on Banach Spaces*,
DOI 10.1007/978-3-319-06447-5_4, © Springer International Publishing Switzerland 2014

which is equivalent to

$$\begin{cases} x_{n+1} - T x_n = f(n, x_n), n \in \mathbb{Z}_+, \\ \\ x_0 = 0, \end{cases} \qquad (4.1.2)$$

where $T := I + A$, $A \in \mathcal{B}(X)$.

To establish the first result, we need to introduce the following assumption:

Assumption (A_1): Suppose that the following conditions hold:

(i) The function $f : \mathbb{Z}_+ \times X \longrightarrow X$ satisfies a Lipschitz condition on X, i.e., for all $z, w \in X$ and $n \in \mathbb{Z}_+$, we have

$$\|f(n, z) - f(n, w)\|_X \leq \alpha_n \|z - w\|_X,$$

where $\alpha := (\alpha_n) \in l_1$.
(ii) $f(\cdot, 0) \in l_1$.

With the above notations we have the following result:

Theorem 4.1.1. *Assume that condition (A_1) holds. In addition suppose that $(3.1.1)$ has discrete maximal l_p-regularity. Then, there is a unique bounded solution $x = (x_n)$ of $(4.1.1)$ such that $(\Delta x) \in l_p(\mathbb{Z}_+; X)$. Moreover, we have the following a priori estimates for the solution*

$$\|x\|_\infty \leq M(T) \|f(\cdot, 0)\|_1 e^{M(T)\|\alpha\|_1}, \qquad (4.1.3)$$

and

$$\|\Delta x\|_p \leq C \|f(\cdot, 0)\|_1 e^{2M(T)\|\alpha\|_1}, \quad 1 < p < +\infty, \qquad (4.1.4)$$

where $M(T) := \sup_{n \in \mathbb{Z}_+} \|T^n\|$ and $C > 0$ is a suitable constant.

Proof. Denote by $\mathcal{W}_0^{\infty, p}$ the Banach space of all sequences $V = (V_n)$ belonging to $l_\infty(\mathbb{Z}_+; X)$ such that $V_0 = 0$ and $\Delta V \in l_p(\mathbb{Z}_+; X)$ equipped with the norm

$$\|V\|_{\mathcal{W}_0^{\infty, p}} = \|V\|_\infty + \|\Delta V\|_p.$$

Let V be a sequence in $\mathcal{W}_0^{\infty, p}$, taking advantage of Assumption (A_1), we first note that the function $g(\cdot) := f(\cdot, V_\cdot)$ is in $l_p(\mathbb{Z}_+; X)$. In fact, we have the following estimates:

$$\|g\|_p^p \leq 2^p \sum_{n=0}^{\infty} \|f(n, V_n) - f(n, 0)\|_X^p + 2^p \sum_{n=0}^{\infty} \|f(n, 0)\|_X^p$$

$$\leq 2^p \sum_{n=0}^{\infty} \alpha_n^p \|V_n\|_X^p + 2^p \sum_{n=0}^{\infty} \|f(n, 0)\|_X^{p-1} \|f(n, 0)\|_X \tag{4.1.5}$$

$$\leq 2^p \|V\|_{\infty}^p \sum_{n=0}^{\infty} \alpha_n^p + 2^p \|f(\cdot, 0)\|_{\infty}^{p-1} \|f(\cdot, 0)\|_1$$

$$\leq 2^p \|V\|_{\infty}^p \|\alpha\|_{\infty}^{p-1} \|\alpha\|_1 + 2^p \|f(\cdot, 0)\|_{\infty}^{p-1} \|f(\cdot, 0)\|_1,$$

which implies that $g \in l_p(\mathbb{Z}_+; X)$.

Since (3.1.1) has discrete maximal l_p-regularity, the Cauchy problem

$$\begin{cases} z_{n+1} - T z_n = g_n, \\ \\ z_0 = 0, \end{cases} \tag{4.1.6}$$

has a unique solution $z = (z_n)$ such that $\Delta z \in l_p(\mathbb{Z}_+; X)$, which is given by

$$z_n = (\mathcal{G}V)_n = \begin{cases} 0 & \text{if } n = 0, \\ \\ \sum_{k=0}^{n-1} T^k f(n-1-k, V_{n-1-k}) & \text{if } n \geq 1. \end{cases} \tag{4.1.7}$$

We now show that the operator $\mathcal{G} : W_0^{\infty,p} \longrightarrow W_0^{\infty,p}$ has a unique fixed point, for this we use the Fixed-point iteration method. Initially we need to verify that $\mathcal{G}$ is well defined. Actually we have only to show that $\mathcal{G}V \in l_{\infty}(\mathbb{Z}_+; X)$. In fact, we use Assumption (A_1) as above to obtain

$$\left\| \sum_{k=0}^{n-1} T^k f(n-1-k, V_{n-1-k}) \right\|_X$$

$$\leq M(T) \sum_{k=0}^{n-1} \alpha_{n-1-k} \|V_{n-1-k}\|_X + M(T) \sum_{j=0}^{n-1} \|f(j, 0)\|_X \tag{4.1.8}$$

$$\leq M(T) \|V\|_{\infty} \sum_{j=0}^{n-1} \alpha_j + M(T) \sum_{j=0}^{n-1} \|f(j, 0)\|_X$$

$$\leq M(T) \Big(\|V\|_{\infty} \|\alpha\|_1 + \|f(\cdot, 0)\|_1 \Big).$$

This proves that the space $\mathcal{W}_0^{\infty,p}$ is invariant under $\mathcal{G}$.

Now, we associate with T the $\mathcal{B}(X)$–valued kernel $k_T : \mathbb{Z} \to \mathcal{B}(X)$ defined by

$$k_T(n) = \begin{cases} (I - T)T^n & \text{for } n \in \mathbb{Z}_+, \\ \\ 0 & \text{otherwise,} \end{cases}$$

and the corresponding operator on $\mathbb{Z}_+$

$$\mathcal{K}_T : l_p(\mathbb{Z}_+; X) \to l_p(\mathbb{Z}_+; X)$$

by

$$(\mathcal{K}_T f)(n) = \sum_{j=0}^{n} k_T(j) f_{n-j}, \quad n \in \mathbb{Z}_+. \tag{4.1.9}$$

By the discrete maximal l_p-regularity, $\mathcal{K}_T$ is well defined and bounded on $l_p(\mathbb{Z}_+; X)$, that is,

$$\|\mathcal{K}_T f\|_p \leq \|\mathcal{K}_T\|_{\mathcal{B}(l_p(\mathbb{Z}_+;X))} \|f\|_p, \text{ for all } f \in l_p(\mathbb{Z}_+; X).$$

Let V and $\tilde{V}$ be in $\mathcal{W}_0^{\infty,p}$. In view of Assumption (A$_1$)-(i) and by using the same computation as in the derivation of (4.1.8), we have initially

$$\|\mathcal{G}V - \mathcal{G}\tilde{V}\|_\infty \leq M(T) \sup_{n \in \mathbb{Z}^+} \left(\sum_{j=0}^{n} \alpha_j \|V_j - \tilde{V}_j\|_X \right) \tag{4.1.10}$$

$$\leq M(T)\|\alpha\|_1 \|V - \tilde{V}\|_\infty.$$

Next we want to estimate $\|\Delta \mathcal{G}V - \Delta \mathcal{G}\tilde{V}\|_p$. Using Minkowski's inequality and taking into account that $\mathcal{K}_T$ is bounded on $l_p(\mathbb{Z}_+, X)$, we can infer the following estimates:

$$\|\Delta \mathcal{G}V - \Delta \mathcal{G}\tilde{V}\|_p$$

$$\leq \left(\sum_{n=1}^{\infty} \|f(n, V_n) - f(n, \tilde{V}_n)\|_X^p \right)^{1/p}$$

$$+ \left(\sum_{n=1}^{\infty} \left\| \sum_{k=0}^{n-1} (T-I)T^k (f(n-1-k, V_{n-1-k}) - f(n-1-k, \tilde{V}_{n-1-k})) \right\|_X^p \right)^{1/p}$$

$$\leq (1 + ||\mathcal{K}_T||_{\mathcal{B}(l_p(\mathbb{Z}_+;X))}) \Big(\sum_{n=1}^{\infty} ||f(n, V_n) - f(n, \tilde{V}_n)||_X^p \Big)^{1/p}$$

$$\leq (1 + ||\mathcal{K}_T||_{\mathcal{B}(l_p(\mathbb{Z}_+;X))}) \Big(\sum_{n=1}^{\infty} \alpha_n^p ||V_n - \tilde{V}_n||_X^p \Big)^{1/p} \qquad (4.1.11)$$

$$\leq (1 + ||\mathcal{K}_T||_{\mathcal{B}(l_p(\mathbb{Z}_+;X))}) ||\alpha||_1 ||V - \tilde{V}||_{\infty}.$$

Hence, summarizing the previous two estimates (4.1.10) and (4.1.11), we obtain

$$||\mathcal{G}V - \mathcal{G}\tilde{V}||_{\mathcal{W}_0^{\infty,p}} \leq M(T) ||\alpha||_1 ||V - \tilde{V}||_{\mathcal{W}_0^{\infty,p}}$$

$$+ (1 + ||\mathcal{K}_T||_{\mathcal{B}(l_p(\mathbb{Z}_+;X))}) ||\alpha||_1 ||V - \tilde{V}||_{\mathcal{W}_0^{\infty,p}}$$

$$= ab ||V - \tilde{V}||_{\mathcal{W}_0^{\infty,p}},$$

where $a := M(T) ||\alpha||_1$ and $b := 1 + (1 + ||\mathcal{K}_T||_{\mathcal{B}(l_p(\mathbb{Z}_+;X))}) M(T)^{-1}$.

Next, we want to calculate the iterates of the operator $\mathcal{G}$. Taking into account Proposition 4.3.1 in Sect. 4.3. We first observe that

$$||[\mathcal{G}^2 V]_n - [\mathcal{G}^2 \tilde{V}]_n||_X \leq M(T) \sum_{j=0}^{n-1} \alpha_j ||[\mathcal{G}V]_j - [\mathcal{G}\tilde{V}]_j||_X$$

$$\leq M(T)^2 \sum_{j=0}^{n-1} \alpha_j \Big(\sum_{i=0}^{j-1} \alpha_i ||V_i - \tilde{V}_i||_X \Big)$$

$$\leq \frac{1}{2} M(T)^2 \Big(\sum_{\tau=0}^{n-1} \alpha_\tau \Big)^2 ||V - \tilde{V}||_{\infty}.$$

Therefore,

$$||\mathcal{G}^2 V - \mathcal{G}^2 \tilde{V}||_{\infty} \leq \frac{1}{2} (M(T) ||\alpha||_1)^2 ||V - \tilde{V}||_{\mathcal{W}_0^{\infty,p}}. \qquad (4.1.12)$$

Furthermore, using (4.1.11) we have

$$\|\Delta\mathcal{G}^2 V - \Delta\mathcal{G}^2\tilde{V}\|_p \leq (1 + \|\mathcal{K}_T\|_{\mathcal{B}(l_p(\mathbb{Z}_+;X))})$$

$$\times\Big(\sum_{n=0}^{\infty}\alpha_n^p\|[\mathcal{G}V]_n - [\mathcal{G}\tilde{V}]_n\|_X^p\Big)^{1/p}$$

$$\leq M(T)(1 + \|\mathcal{K}_T\|_{\mathcal{B}(l_p(\mathbb{Z}_+;X))})$$

$$\times\Big(\sum_{n=0}^{\infty}\alpha_n^p\Big(\sum_{k=0}^{n-1}\alpha_k\|V_k - \tilde{V}_k\|_X\Big)^p\Big)^{1/p}$$

$$\leq M(T)(1 + \|\mathcal{K}_T\|_{\mathcal{B}(l_p(\mathbb{Z}_+;X))})$$

$$\times\Big(\sum_{n=0}^{\infty}\alpha_n^p\Big(\sum_{j=0}^{n-1}\alpha_j\Big)^p\|V - \tilde{V}\|_{\infty}^p\Big)^{1/p}$$

$$\leq \frac{1}{2}M(T)(1 + \|\mathcal{K}_T\|_{\mathcal{B}(l_p(\mathbb{Z}_+;X))})$$

$$\times\Big(\sum_{j=0}^{\infty}\alpha_j\Big)^2\|V - \tilde{V}\|_{W_0^{\infty,p}},$$

whence

$$\|\mathcal{G}^2 V - \mathcal{G}^2\tilde{V}\|_{W_0^{\infty,p}} \leq \frac{b}{2}a^2\|V - \tilde{V}\|_{W_0^{\infty,p}}, \tag{4.1.13}$$

with a and b defined as above. Applying the above estimates successively an induction argument shows us that

$$\|\mathcal{G}^n V - \mathcal{G}^n\tilde{V}\|_{W_0^{\infty,p}} \leq \frac{b}{n!}a^n\|V - \tilde{V}\|_{W_0^{\infty,p}}. \tag{4.1.14}$$

Since $ba^n/n! < 1$ for n sufficiently large, by the fixed-point iteration method, $\mathcal{G}$ has a unique fixed point $V \in W_0^{\infty,p}$. Let V be the unique fixed point of $\mathcal{G}$, then by Assumption (A_1) we have

$$\|V_{n+1}\|_X \leq M(T)\sum_{j=0}^{n}\alpha_j\|V_j\|_X + M(T)\sum_{j=0}^{n}\|f(j,0)\|_X$$

$$\tag{4.1.15}$$

$$\leq M(T)\sum_{j=0}^{n}\alpha_j\|V_j\|_X + M(T)\|f(\cdot,0)\|_1.$$

Now, as an application of the discrete Gronwall's inequality[1] [1, Corollary 4.12, p. 183], we get immediately

$$||V||_\infty \le M(T)||f(\cdot, 0)||_1 e^{M(T)||\alpha||_1}. \tag{4.1.16}$$

Finally we estimate $||\Delta V||_p$, using the fact that $\Delta V_0 = f(0, 0)$, and proceeding analogously as in (4.1.11), we deduce the following estimates:

$$||\Delta V||_p \le ||f(0, 0)||_X + \left(\sum_{n=1}^\infty ||\Delta V_n||_X^p \right)^{1/p}$$

$$\le ||f(0, 0)||_X + \left(\sum_{n=1}^\infty ||f(n, V_n)||_X^p \right)^{1/p}$$

$$+ ||\mathcal{K}_T||_{\mathcal{B}(l_p(\mathbb{Z}_+; X))} \left(\sum_{n=0}^\infty ||f(n, V_n)||_X^p \right)^{1/p}$$

$$\le (2 + ||\mathcal{K}_T||_{\mathcal{B}(l_p(\mathbb{Z}_+; X))}) \sum_{n=0}^\infty ||f(n, V_n)||_X,$$

where, by Assumption (A_1) and (4.1.16), we infer that

$$\sum_{n=0}^\infty ||f(n, V_n)||_X \le \sum_{n=0}^\infty \alpha_n ||V_n||_X + ||f(\cdot, 0)||_1$$

$$\le ||\alpha||_1 ||V||_\infty + ||f(\cdot, 0)||_1$$

$$\le ||\alpha||_1 M(T)||f(\cdot, 0)||_1 e^{M(T)||\alpha||_1} + ||f(\cdot, 0)||_1$$

$$\le ||f(\cdot, 0)||_1 e^{2M(T)||\alpha||_1}.$$

This ends the proof of the theorem.   □

---

[1]Gronwall's inequality is one method in which one can assure the absence of blowup. The finite difference version of the well-known Gronwall's inequality seems to have appeared first in the work of Mikeladze [145] in 1935. It is well recognized that the discrete version of Gronwall's inequality provides a very useful and important tool in proving convergence of discrete variable methods. A detailed exposition of the inequalities of Gronwall discrete type is contained in Agarwal's book [1]. It is noted that some difference inequalities obtained for difference equations can be applied to the stability and boundedness problems for differential and functional-differential equations as also can be applied to the error estimation in the numerical analysis. Owing to the considerable applications, recently some new finite difference inequalities are developed to widen the scope of their applications (see [155]).

In view of Blunck's theorem (see Theorem 3.1.4), we obtain the following result which is valid on UMD spaces.

Corollary 4.1.2. *Let X be a UMD space. Assume that condition (A_1) holds and suppose $T \in \mathcal{B}(X)$ is power bounded and analytic such that the set*

$$\left\{ (z - 1)\left(z - T\right)^{-1} : |z| = 1, z \neq 1 \right\},$$

is R-bounded. Then, there is a unique bounded solution $x = (x_n)$ of (4.1.1) such that $(\Delta x) \in l_p(\mathbb{Z}_+; X)$. Moreover, the a priori estimates (4.1.3) and (4.1.4) hold.

Example 4.1.3. Consider the semilinear problem

$$\Delta x_n - (T - I)x_n = q_n f(x_n), \tag{4.1.17}$$

where f is defined and satisfies a Lipschitz condition with constant L on a Hilbert space H. In addition suppose $(q_n) \in l_1$. Then Assumption (A_1) is satisfied. In our case, applying the preceding result, we obtain that if $T \in \mathcal{B}(H)$ is power bounded and analytic and satisfies that the set

$$\left\{ (\lambda - 1)\left(\lambda - T\right)^{-1} : |\lambda| = 1, \lambda \neq 1 \right\}$$

is bounded, then there is a unique bounded solution $x = (x_n)$ of the (4.1.17) such that $(\Delta x) \in l_p(\mathbb{Z}_+; H)$. Moreover,

$$\|x\|_\infty \leq M(T)\|f(0)\|_H \|q\|_1 e^{LM(T)\|q\|_1}. \tag{4.1.18}$$

In particular, taking $T = I$ the identity operator and $H = \mathbb{C}$, we obtain the following result.

Corollary 4.1.4. *Suppose f is defined and satisfies a Lipschitz condition with constant L. Let $(q_n) \in l_1(\mathbb{Z}_+)$; then the equation*

$$\Delta x_n = q_n f(x_n) \tag{4.1.19}$$

has a unique bounded solution $x = (x_n)$ such that $(\Delta x_n) \in l_p(\mathbb{Z}_+; \mathbb{C})$ and (4.1.18) holds with $M(T) = 1$.

We remark that the above result covers a wide range of difference equations.

4.2 Local Perturbations

Note that Theorem 4.1.1 is not general enough to include perturbations like the following:

$$f(n, z) = y_n + \alpha_n B(z, z), \tag{4.2.1}$$

where $B : X \times X \to X$ is a bounded bilinear operator and α_n, y_n are fixed sequences. In fact, it does not satisfy condition (A_1)-(i). This leads to us to study locally Lipschitzian perturbations of (4.1.1).

In the process of obtaining our next result, we will require the following assumption.

Assumption $(A_1)^*$: The following conditions hold:

(i)* The function $f(n, z)$ is locally Lipschitz with respect to $z \in X$, i.e., for each positive number R, for all $n \in \mathbb{Z}_+$, and $z, w \in X$, $||z||_X \leq R$, $||w||_X \leq R$, we have

$$||f(n, z) - f(n, w)||_X \leq l(n, R)||z - w||_X,$$

where $\ell : \mathbb{Z}_+ \times [0, \infty) \longrightarrow [0, \infty)$ is a nondecreasing function with respect to the second variable.

(ii)* There is a positive number a such that

$$\sum_{n=0}^{\infty} \ell(n, a) < +\infty.$$

(iii)* $f(\cdot, 0) \in \ell_1$.

Our main result in this section is the following local version of Theorem 4.1.1.

Theorem 4.2.1. *Suppose that the following conditions are satisfied:*

(a)* *The condition $(A_1)^*$ holds.*
(b)* *Equation (3.1.1) has maximal l_p-regularity.*

Then, there is a positive constant $m \in \mathbb{N}$ and a unique bounded solution $x = (x_n)$ of (4.1.1) for $n > m$ such that $x_n = 0$ if $0 \leq n \leq m$ and the sequence (Δx_n) belongs to $\ell_p(\mathbb{Z}_+; X)$. Moreover, we get

$$||x||_\infty + ||\Delta x||_p \leq a, \tag{4.2.2}$$

where a is the constant of condition $(A_1)^*$-(ii).*

Proof. Let $\beta \in (0, 1)$. Using (iii)* and (ii)* there are n_1 and n_2 in $\mathbb{N}$ such that

$$(M(T) + 2 + ||\mathcal{K}_T||_{\mathcal{B}(l_p(\mathbb{Z}_+;X))}) \sum_{j=n_1}^{\infty} ||f(j, 0)||_X \leq \beta a, \tag{4.2.3}$$

and

$$T := \beta + (M(T) + 1 + ||\mathcal{K}_T||_{\mathcal{B}(l_p(\mathbb{Z}_+;X))}) \sum_{j=n_2}^{\infty} \ell(j,a) < 1. \tag{4.2.4}$$

Putting $m = \max\{n_1, n_2\}$. We denote by $\mathcal{W}_m^{\infty,p}$ the Banach space[2] of all sequences $V = (V_n)$ belonging to $\ell_\infty(\mathbb{Z}_+; X)$ such that $V_n = 0$ if $0 \le n \le m$, and $\Delta V \in \ell_p(\mathbb{Z}_+; X)$ equipped with the norm $||\cdot||_{\mathcal{W}_0^{\infty,p}}$ given in the Theorem 4.1.1. We denote by $\mathcal{W}_m^{\infty,p}[a]$ the closed ball $||V||_{\mathcal{W}_0^{\infty,p}} \le a$ in $\mathcal{W}_m^{\infty,p}$.

Let V be a sequence in $\mathcal{W}_m^{\infty,p}[a]$. A short argument similar to (4.1.5) and Assumption $(A_1)^*$ shows that the sequence

$$g_n := \begin{cases} 0 & \text{if } 0 \le n \le m, \\ \\ f(n, V_n) & \text{if } n > m, \end{cases} \tag{4.2.5}$$

belongs to ℓ_p. By the discrete maximal l_p-regularity, the Cauchy's problem (4.1.6) with g_n defined as in (4.2.5) has a unique solution (z_n) such that

$$\Delta z_n \in l_p(\mathbb{Z}_+; X),$$

which is given by

$$z_n = [\mathcal{D}V]_n = \begin{cases} 0 & \text{if } 0 \le n \le m, \\ \\ \sum_{k=0}^{n-1-m} T^k f(n-1-k, V_{n-1-k}) & \text{if } n \ge m+1. \end{cases} \tag{4.2.6}$$

---

[2]Let $(V^l)_l$ be a Cauchy sequence in $\mathcal{W}_m^{\infty,p}$; then $(V^l)_l$ and $(\Delta V^l)_l$ are Cauchy sequences in $\ell_\infty(\mathbb{Z}_+; X)$ and $\ell_p(\mathbb{Z}_+; X)$, respectively. Therefore there are sequences $V \in \ell_\infty(\mathbb{Z}_+; X)$ and $Y \in \ell_p(\mathbb{Z}_+; X)$ such that

$$V^l \to V, \text{ in } \ell_\infty(\mathbb{Z}_+; X) \text{ as } l \to \infty,$$

$$\Delta V^l \to Y, \text{ in } \ell_p(\mathbb{Z}_+; X) \text{ as } l \to \infty.$$

We observe that

$$||\Delta V_k^l - Y_k||_X \le ||\Delta V^l - Y||_p,$$

$$||V_k^l - Y_k||_X \le ||V^l - Y||_\infty.$$

Letting $l \to \infty$, we obtain that $Y = \Delta V$ and $V_k = 0$ for $0 \le k \le m$. Therefore $V \in \mathcal{W}_m^{\infty,p}$ and $V^l \to V$ in $\mathcal{W}_m^{\infty,p}$ as $l \to \infty$.

We will prove that $\mathcal{D}V$ belongs to $\mathcal{W}_m^{\infty,p}[a]$. In fact, since

$$||V_j||_X \leq ||V||_\infty$$
$$\leq ||V||_{\mathcal{W}_0^{\infty,p}}$$
$$< a,$$

we have by Assumption $(A_1)^*$

$$||[\mathcal{D}V]_n||_X \leq M(T) \sum_{j=m}^{n-1} l(j,a)||V_j||_X + M(T) \sum_{j=m}^{n-1} ||f(j,0)||_X$$

$$\tag{4.2.7}$$

$$\leq M(T) \sum_{j=m}^{\infty} l(j,a)a + M(T) \sum_{j=m}^{\infty} ||f(j,0)||_X.$$

Using that $V_m = 0$, we can proceed analogously as in (4.1.11) to obtain

$$||\Delta \mathcal{D}V||_p = \left(||f(m,0)||_X^p + \sum_{n=m+1}^{\infty} ||\mathcal{D}V_{n+1} - \mathcal{D}V_n||_X^p \right)^{1/p}$$

$$\leq ||f(m,0)||_X + (1 + ||\mathcal{K}_T||_{\mathcal{B}(l_p(\mathbb{Z}_+;X))})$$

$$\times \left(\sum_{n=m+1}^{\infty} ||f(n,V_n)||_X^p \right)^{1/p}$$

$$\leq \sum_{n=m}^{\infty} ||f(n,0)||_X + (1 + ||\mathcal{K}_T||_{\mathcal{B}(l_p(\mathbb{Z}_+;X))}) \sum_{n=m}^{\infty} ||f(n,V_n)||_X.$$

Therefore using (4.2.7) we get

$$||\Delta \mathcal{D}V||_p \leq \sum_{j=m}^{\infty} ||f(j,0)||_X + (1 + ||\mathcal{K}_T||_{\mathcal{B}(l_p(\mathbb{Z}_+;X))})$$

$$\times \left(\sum_{j=m}^{\infty} l(j,a)a + \sum_{j=m}^{\infty} ||f(j,0)||_X \right)$$

$$= (1 + ||\mathcal{K}_T||_{\mathcal{B}(l_p(\mathbb{Z}_+;X))}) \sum_{j=m}^{\infty} l(j,a)a + (2 + ||\mathcal{K}_T||_{\mathcal{B}(l_p(\mathbb{Z}_+;X))})$$

$$\times \sum_{j=m}^{\infty} ||f(j,0)||_X.$$

$$\tag{4.2.8}$$

Then, inequalities (4.2.7) and (4.2.8) together with (4.2.3) and (4.2.4) imply

$$||\mathcal{D}V||_{\mathcal{W}_0^{\infty,p}} \leq (M(T) + 1 + ||\mathcal{K}_T||_{\mathcal{B}(l_p(\mathbb{Z}_+;X))}) \sum_{j=m}^{\infty} \ell(j,a)a$$

$$+ (M(T) + 2 + ||\mathcal{K}_T||_{\mathcal{B}(l_p(\mathbb{Z}_+;X))}) \sum_{j=m}^{\infty} ||f(j,0)||_X$$

$$\leq (1 - \beta)a + \beta a = a,$$

proving that $(\mathcal{D}V)$ belongs to $\mathcal{W}_m^{\infty,p}[a]$.

Furthermore, for all V and W in $\mathcal{W}_m^{\infty,p}[a]$, we have

$$||\mathcal{D}V - \mathcal{D}W||_{\mathcal{W}_0^{\infty,p}} \leq M \sum_{j=m}^{\infty} \ell(j,a)||V - W||_{\mathcal{W}_0^{\infty,p}} + (1 + ||\mathcal{K}_T||_{\mathcal{B}(l_p(\mathbb{Z}_+;X))})$$

$$\times \sum_{j=m}^{\infty} \ell(j,a)||V - W||_{\mathcal{W}_0^{\infty,p}}$$

$$= (\mathcal{T} - \beta)||V - W||_{\mathcal{W}_0^{\infty,p}}.$$

Hence $\mathcal{D}$ is a $(\mathcal{T} - \beta)$-contraction. This completes the proof of the theorem. □

This enable us to prove, as an application, the following corollary.

Corollary 4.2.2. *Let $B : X \times X \longrightarrow X$ be a bounded, bilinear Operator, $y \in \ell_1(\mathbb{Z}_+;X)$, and $\alpha \in \ell_1(\mathbb{Z}_+;\mathbb{R})$. In addition suppose that $T \in \mathcal{B}(X)$ is power bounded and analytic such that the set*

$$\left\{ (z - 1)(z - T)^{-1} : |z| = 1, z \neq 1 \right\},$$

is R-bounded. Then, there is a unique bounded solution $x = (x_n)$ such that $(\Delta x) \in l_p(\mathbb{Z}_+;X)$ for the equation

$$x_{n+1} - Tx_n = y_n + \alpha_n B(x_n, x_n).$$

In the particular case $T = I$ and $X = M(n \times n)$ the set of all $n \times n$ matrices, we obtain for $B(M, N) = MN$ the following result:

Corollary 4.2.3. *Let $Y \in \ell_1(\mathbb{Z}_+; M(n \times n))$ and $\alpha \in \ell_1(\mathbb{Z}_+)$. Then, there is a unique bounded solution Z such that*

$$(\Delta Z) \in l_p(\mathbb{Z}_+; M(n \times n))$$

for the equation

$$Z_{n+1} - Z_n = Y_n + \alpha_n Z_n^2.$$

4.3 Comments

The following proposition is very useful to calculate the iterates of the operator $\mathcal{G}$ in the proof of Theorem 4.1.1.

Proposition 4.3.1. *Let* $(\alpha_m)_m$ *be a sequence of positive real numbers. For all* $n, l \in \mathbb{Z}_+$, *we have*[3]

$$\sum_{m=0}^{n-1} \alpha_m \Big(\sum_{j=0}^{m-1} \alpha_j \Big)^l \leq \frac{1}{l+1} \Big(\sum_{j=0}^{n-1} \alpha_j \Big)^{l+1}.$$

Theorems 4.1.1 and 4.2.1 are taken from [54].

The methods presented in this section can be used to study the existence and uniqueness of bounded solutions which are in l_p for semilinear functional difference equations with infinite delay. The literature concerning discrete maximal regularity for functional difference equations with infinite delay is too incipient and should be developed, so as to produce a significant progress in the theory of abstract functional difference equations; see Sect. 7.7.

---

[3]Putting $A_m := \sum_{j=0}^{m-1} \alpha_j$, we obtain

$$(l+1)(A_{m+1} - A_m)A_m^l = (A_{m+1} - A_m)(A_m^l + A_m^{l-1}A_m + \ldots + A_m A_m^{l-1} + A_m^l)$$

$$\leq (A_{m+1} - A_m)(A_{m+1}^l + A_{m+1}^{l-1}A_m + \ldots + A_{m+1}A_m^{l-1} + A_m^l)$$

$$= A_{m+1}^{l+1} - A_m^{l+1}.$$

Hence

$$\sum_{m=0}^{n-1}(A_{m+1} - A_m)A_m^l \leq \frac{1}{l+1}\sum_{m=0}^{n-1}(A_{m+1}^{l+1} - A_m^{l+1}) = \frac{1}{l+1}A_n^{l+1}.$$

Chapter 5
Second-Order Linear Difference Equations

This chapter introduces the notion of discrete maximal regularity for second-order linear difference equations. In analogy to the case of first-order linear difference equations studied previously, we obtain a characterization of maximal l_p-regularity. The study of reduction of order is also treated in this chapter. We also provide examples and several comments concerning open problems on this part of the theory.

5.1 Discrete Maximal Regularity

Let A be a bounded linear operator on X. For a given sequence $(f_n)_{n \in \mathbb{Z}_+}$ in X, we consider in this section the problem of producing a sequence $(x_n)_{n \in \mathbb{Z}_+}$ such that

$$\Delta^2 x_n - (I - T)x_n = f_n, \qquad (5.1.1)$$

for all $n \in \mathbb{Z}_+$ with the initial conditions $x_0 = 0$ and $x_1 = 0$.

We associate with T the operator

$$\mathcal{K}_\mathcal{S} f(n) := (I - T)\mathcal{S} * f(n) = \sum_{k=1}^{n} (I - T)\mathcal{S}(k) f_{n-k}, \qquad (5.1.2)$$

where $\mathcal{S}(\cdot)$ is the discrete sine operator (see Sect. 1.4. Note that in [53] the sine operator is given explicitly).

The following definition is the natural extension of the concept of maximal regularity for the continuous case; cf. [156].

Definition 5.1.1. Let $1 < p < +\infty$. We say that (5.1.1) has discrete maximal l_p-regularity if $\mathcal{K}_\mathcal{S} \in \mathcal{B}(l_p(\mathbb{Z}_+; X))$.

R.P. Agarwal et al., *Regularity of Difference Equations on Banach Spaces*,
DOI 10.1007/978-3-319-06447-5_5, © Springer International Publishing Switzerland 2014

As a consequence of this definition and Proposition 1.4.5, if (5.1.1) has discrete maximal l_p-regularity, then for each $(f_n) \in l_p(\mathbb{Z}_+; X)$ we have

$$\Delta^2 x_{n+1} = f_{n+1} + (I - T)x_{n+1}$$

$$= f_{n+1} + ((I - T)\mathcal{S} * f)(n)$$

$$= f_{n+1} + \mathcal{K}_\mathcal{S} f(n) \in l_p(\mathbb{Z}_+; X).$$

The following is the counterpart of Blunck's result for second-order linear difference equations. It was proved in [53].

Theorem 5.1.2. *Let X be a UMD space and let $T \in \mathcal{B}(X)$ be analytic. Then the following assertions are equivalent:*

(i) *Equation (5.1.1) has discrete maximal l_p-regularity.*

(ii) *The set $\left\{ (z - 1)^2 \big((z-1)^2 - (I - T) \big)^{-1} : |z| = 1, z \neq 1 \right\}$ is R-bounded.*

Proof. Since T is analytic, we deduce[1] that $((z - 1)^2 - (I - T))^{-1}$ exists for $|z| = 1, z \neq 1$. (ii) $\implies$ (i). By definition, (5.1.1) has discrete maximal l_p-regularity if and only if the operator

$$\mathcal{K}_\mathcal{S} f(n) := \sum_{k=0}^{n} (I - T)\mathcal{S}(k) f_{n-k}$$

is bounded on $l_p(\mathbb{Z}_+; X)$. Define $\mathcal{K}_\mathcal{S} f(n)$ by 0 for $n < 0$. Taking the time-discrete Fourier transform, we obtain by Proposition 1.4.2

$$\widehat{\mathcal{K}_\mathcal{S} f}(z) = (I - T)z\big((z - 1)^2 - (I - T) \big)^{-1} \hat{f}(z). \qquad (5.1.3)$$

From the identity

$$((z - 1)^2 - (I - T)) R((z - 1)^2, I - T) = I,$$

we get

$$(I - T)\big((z - 1)^2 - (I - T) \big)^{-1} = (z - 1)^2 \big((z - 1)^2 - (I - T) \big)^{-1} - I$$

then multiplying by z, we obtain

---

[1] We observe that $\sigma(I - T) \subseteq D(1, 1) \cup \{0\}$, hence $(z - 1)^2 \in \rho(I - T)$ whenever $|z| = 1, z \neq 1$.

$$(I - T)z\big((z - 1)^2 - (I - T)\big)^{-1} = z\Big((z - 1)^2\big((z - 1)^2 - (I - T)\big)^{-1} - I\Big).$$

Hence

$$\widehat{\mathcal{K}_{\mathcal{S}} f}(z) = z\Big((z - 1)^2\big((z - 1)^2 - (I - T)\big)^{-1} - I\Big)\hat{f}(z). \qquad (5.1.4)$$

By hypothesis, the set

$$\Big\{(z - 1)^2\big((z - 1)^2 - (I - T)\big)^{-1} : |z| = 1, \ z \neq 1\Big\}$$

is R-bounded. Define

$$M(t) := (e^{it} - 1)^2\big((e^{it} - 1)^2 - (I - T)\big)^{-1}. \qquad (5.1.5)$$

Then

$$(e^{it} - 1)(e^{it} + 1)M'(t) = 2i\,e^{it}(e^{it} + 1)M(t) - 2i\,e^{it}(e^{it} + 1)(M(t))^2, \quad t \in (0, 2\pi). \qquad (5.1.6)$$

We conclude by hypothesis and the permanence properties of R-bounded sets (Proposition 2.2.5) that the sets

$$\{M(t) : t \in (0, 2\pi)\} \text{ and } \{(e^{it} - 1)(e^{it} + 1)M'(t) : t \in (0, 2\pi)\}$$

are R-bounded.

Then, by Blunck's theorem (Theorem 2.4.9), there is an operator $T_M \in \mathcal{B}(l_p(\mathbb{Z}; X))$ such that

$$\widehat{T_M f}(z) = (z - 1)^2\big((z - 1)^2 - (I - T)\big)^{-1}\hat{f}(z).$$

Hence (5.1.4) implies that

$$\widehat{\mathcal{K}_{\mathcal{S}} f}(z) = z\widehat{T_M f}(z) - z\hat{f}(z).$$

Then, by uniqueness of the Z-transform, $\mathcal{K}_{\mathcal{S}}$ is bounded on $l_p(\mathbb{Z}_+; X)$, proving (i).

(i) $\implies$ (ii). Define

$$Kf(n) := (\mathcal{K}_{\mathcal{S}} f)(n - 1) + (\delta * f)(n), \qquad (5.1.7)$$

where $\delta[n]$ denotes the Kronecker delta ($= 1$ for $n = 0$ and zero otherwise). Then K is a bounded operator in $l_p(\mathbb{Z}; X)$ for which, in view of (5.1.4), we have

$$\widehat{Kf}(z) = z^{-1}\widehat{\mathcal{K}_{\mathcal{S}}f}(z) + \hat{f}(z) = (z-1)^2\Big((z-1)^2 - (I-T)\Big)^{-1}\hat{f}(z). \quad (5.1.8)$$

It follows from Proposition 2.5.11 that the set

$$\left\{(z-1)^2\Big((z-1)^2 - (I-T)\Big)^{-1} : |z| = 1, z \neq 1\right\}$$

is R-bounded.   □

In case of Hilbert spaces, we deduce the following result.

Corollary 5.1.3. *Let H be a Hilbert space and let $T \in \mathcal{B}(H)$ be analytic. Then the following assertions are equivalent:*

(i) *Equation (5.1.1) has discrete maximal l_p-regularity.*

(ii) *The set $\left\{(z-1)^2\Big((z-1)^2 - (I-T)\Big)^{-1} : |z| = 1, z \neq 1\right\}$ is bounded.*

5.2 Exact Discretizations

In numerical integration of a differential equation a standard approach is to replace it by a suitable difference equation whose solution can be obtained in a stable manner and without troubles from round off errors. However, often the qualitative properties of the solutions of the difference equation are quite different from the solutions of the corresponding differential equations.

For a given differential equation a difference equation approximation is called *exact* (or best) if the solution of the difference equation exactly coincides with solutions of the corresponding differential equation evaluated at a discrete sequence of points. Exact approximations are not unique (cf. [1, Sect. 3.6]).

Exact discretizations have been first studied by Potts [163] and a detailed account of subsequent developments can be found in Agarwal's book [1]. It is worthwhile to point out that all linear ordinary differential equations with constant coefficients admit exact discretizations ([1, 163]).

In 2006, Cieśliński and Ratkiewicz [42] studied various discretizations of the harmonic oscillator equation $\ddot{u}(t) = -\omega^2 u(t)$ and compared them. An exact discretization is given by

$$\Delta^2 x_n + (2\sin\epsilon\omega/2)^2 x_{n+1} = 0. \quad (5.2.1)$$

In other words, any solution x_n of (5.2.1) can be expressed as $x_n = u(\epsilon n)$ (the time step: $t_{n+1} - t_n = \epsilon$ is constant). We note that in the limit $\epsilon \to 0$ the (5.2.1) assumes the form of the symmetric Euler finite difference scheme (see [43, 44] for a discussion on this subject).

Motivated by this result, we study in this section the discrete second-order equation

$$\Delta^2 x_n + A x_{n+1} = f_n, \tag{5.2.2}$$

on complex Banach spaces, where $A \in \mathcal{B}(X)$. Of course, in the finite dimensional setting, (5.2.2) includes systems of linear difference equations, but the most interesting application concerns with partial difference equations. In fact, the homogeneous equation associated to (5.2.2) corresponds to the exact discretization of the wave equation (cf. [1, Sect. 3.14]).

We now give a geometrical link between the best discretization (5.2.2) and the equations of the form

$$\Delta^2 x_n + A x_{n+k} = f_n, \quad x_0 = x_1 = 0, \quad k \in \{0, 1, 2\}. \tag{5.2.3}$$

The motivation comes again from the recent article of Cieśliński and Ratkiewicz [42], where several discretizations of second-order linear ordinary differential equations with constant coefficients are compared and discussed.

Remark 5.2.1. Observe that (5.2.3) can be rewritten as

$$x_{n+2} = 2x_{n+1} - x_n - A x_{n+k} + f_n. \tag{5.2.4}$$

If $k \in \mathbb{Z}$ in (5.2.3), then we have a well-defined recurrence relation of order 2 in case $k = 0$ or 1 (and of order $(2 - k)$) in case $k < 0$. In case $k = 2$, we have

$$(I + A)x_{n+2} = 2x_{n+1} - x_n + f_n,$$

that is a recurrence relation of order 2, which need not be well defined unless $-1 \in \rho(A)$. Finally, in case $k > 2$,

$$x_{n+k} = A^{-1}(2x_{n+1} - x_n - x_{n+2} + f_n)$$

is of order k (note that here we need $0 \in \rho(A)$).

Taking (formally) Z-transform of (5.2.3) we obtain

$$(z - 1)^2 \hat{x}(z) + A z^k \hat{x}(z) = \hat{f}(z).$$

Hence the operator $(z - 1)^2 + z^k A$ is invertible if and only if $\dfrac{-(z-1)^2}{z^k}$ belongs to the resolvent set $\rho(A)$ of A. Define the function

$$\Gamma_\alpha(t) = -\frac{(e^{it} - 1)^2}{e^{i\alpha t}}, \quad \alpha \in \mathbb{R},\ t \in (0, 2\pi). \tag{5.2.5}$$

Then, for each α fixed, $\Gamma_\alpha(t)$ describes a curve in the complex plane such that

$$\Gamma_\alpha(0) = \Gamma_\alpha(2\pi) = 0.$$

Proposition 5.2.2. *The curve Γ_α attains the minimum length at $\alpha = 1$.*

Proof. A calculation gives

$$\Gamma_\alpha'(t) = -2i\,e^{-i\frac{\alpha}{2}t}((\alpha - 1)(1 - \cos t) + i\,\sin t).$$

Hence the length of Γ_α is given by

$$l(\alpha) = \int_0^{2\pi} |\Gamma_\alpha'(t)|\,dt = 2\int_0^{2\pi} \sqrt{(\alpha - 1)^2(1 - \cos t)^2 + \sin^2 t}\,\,dt.$$

$\square$

Remark 5.2.3. As a consequence, the value $k = 1$ in (5.2.3) is *singular* in the sense that the curve described by (5.2.5) attains the minimum length if and only if $\alpha = 1$. This singular character is reinforced by observing that

$$\Gamma_1(\epsilon\omega) = (2\sin\frac{\epsilon\omega}{2})^2,$$

and that this value exactly corresponds to the step size in the best discretization of the harmonic oscillator[2].

Recall the notation

$$T := A + I;\; \mathbb{D}(z, r) = \{w \in \mathbb{C} : |w - z| < r\} \text{ and } \mathbb{T} = \partial\mathbb{D}(0, 1).$$

The following result relates the values of $\Gamma_1(t)$ with the spectrum of the operator A. It will be essential in the proof of our characterization of well-posedness for (5.2.2) in l_p-vector-valued spaces given in the next section.

Proposition 5.2.4. *Suppose that T is analytic. Then $\sigma(I - T) \subseteq \mathbb{D}(1, 1) \cup \{0\}$. In particular,*

$$-\Gamma_1((0, 2\pi)) \subset \rho(I - T).$$

Proof. Let $M > 0$ be such that

$$\frac{M}{n} \geq ||T^n(T - I)||$$

---

[2]We conjecture that there is a general link between the geometrical properties of curves related to classes of difference equations and the property of exact discretization.

for all $n \in \mathbb{N}$. Define $p(z) = z^{n+1} - z^n$. By the spectral mapping theorem, we have

$$\|T^n(T - I)\| \geq \sup_{\lambda \in \sigma(p(T))} |\lambda|$$

$$= \sup_{\lambda \in p(\sigma(T))} |\lambda|$$

$$= \sup_{w \in \sigma(I-T)} |w(1 - w)|^n$$

$$\geq |w||1 - w|^n,$$

for all $w \in \sigma(I - T), n \in \mathbb{N}$. Hence

$$\sigma(I - T) \subseteq \mathbb{D}(1, 1) \cup \{0\}.$$

Finally, we observe that

$$-\Gamma_1(t) = -[2\sin(t/2)]^2 \in (-4, 0), \text{ for } t \in (0, 2\pi).$$

$\square$

5.3   Exact Second-Order Difference Equation

In this section, we treat the existence and uniqueness problem for the following difference equation:

$$\begin{cases} \Delta^2 x_n - (I - T)x_{n+1} = f_n, n \in \mathbb{Z}_+, \\ \\ x_0 = x, \ x_1 = y. \end{cases} \tag{5.3.1}$$

The motivation for the study of this equation was given in Sect. 5.2. Suppose now that T is an analytic operator, such that $1 \in \rho(T)$. Then by Proposition 5.2.4 we have

$$B(n)x := \frac{1}{2\pi} \int_{-\pi}^{\pi} e^{int} \left(\frac{(e^{it} - 1)^2}{e^{it}} - (I - T) \right)^{-1} x \, dt, \quad x \in X, \tag{5.3.2}$$

is a well-defined operator in $\mathcal{B}(X)$. Hence, the discrete-time Fourier transform of $B(n)$ is given by

$$\hat{B}(z) = \left(\frac{(z - 1)^2}{z} - (I - T) \right)^{-1}. \tag{5.3.3}$$

Note that

$$B(n)x = \frac{1}{2\pi i} \int_{|z|=1} z^{n-1} \left(\frac{(z-1)^2}{z} - (I-T) \right)^{-1} x \, dz, \quad x \in X. \qquad (5.3.4)$$

In particular, from Cauchy's integral formula, we deduce that $B(0) = 0$.

Our main result in this section, on existence and uniqueness of solution for (5.3.1), reads as follows.

Theorem 5.3.1. *Let $T \in B(X)$ be analytic such that $1 \in \rho(T)$, then there is a unique solution of (5.3.1) which is given by*

$$x(n+1) = -B(n)x(0) + B(n+1)x(1) + (B * f)(n). \qquad (5.3.5)$$

Proof. We proceed by applying discrete-time Fourier transform to (5.3.1). We directly obtain

$$\hat{x}(z) = \left((z-1)^2 - z(I-T) \right)^{-1} \left(z^2 - 3z + zT \right) x(0)$$

$$+ z\left((z-1)^2 - z(I-T) \right)^{-1} x(1) + \left((z-1)^2 - z(I-T) \right)^{-1} \hat{f}(z).$$

Using now the identity

$$I = \left((z-1)^2 - z(I-T) \right)^{-1} \left((z-1)^2 - z(I-T) \right)$$

$$= \left((z-1)^2 - z(I-T) \right)^{-1} \left(z^2 - 3z + zT \right) + \left((z-1)^2 - z(I-T) \right)^{-1},$$

we have

$$\hat{x}(z) = x(0) - \left((z-1)^2 - z(I-T) \right)^{-1} x(0)$$

$$+ z\left((z-1)^2 - z(I-T) \right)^{-1} x(1)$$

$$+ \left((z-1)^2 - z(I-T) \right)^{-1} \hat{f}(z).$$

Multiplying the above by z we obtain the equivalent identity

$$z\hat{x}(z) - zx(0) = -\left(\frac{(z-1)^2}{z} - (I - T)\right)^{-1} x(0)$$

$$+ z\left(\frac{(z-1)^2}{z} - (I - T)\right)^{-1} x(1)$$

$$+ \left(\frac{(z-1)^2}{z} - (I - T)\right)^{-1} \hat{f}(z),$$

and the result follows by inversion, because $B(0) = 0$. $\qquad\square$

Notice that the above result in a more general setting and without the assumption $1 \in \rho(T)$ appeared in author's work [38].

5.4 A Characterization of Maximal l_p-Regularity

In this section, we obtain a spectral characterization about maximal regularity for the equation

$$\begin{cases} \Delta^2 x_n - (I - T)x_{n+1} = f_n, n \in \mathbb{Z}_+, \\ \\ x_0 = x_1 = 0. \end{cases} \tag{5.4.1}$$

The following definition is motivated by the previous sections.

Definition 5.4.1. Let $1 < p < +\infty$. We say that (5.4.1) has discrete maximal l_p-regularity if $\mathcal{K}_B f := (I - T)B * f$ defines a bounded operator $\mathcal{K}_B \in \mathcal{B}(l_p(\mathbb{Z}_+; X))$.

As a consequence of the definition, if (5.4.1) has discrete maximal l_p-regularity, then for each $(f_n) \in l_p(\mathbb{Z}_+; X)$ we have $(\Delta^2 x_n) \in l_p(\mathbb{Z}_+; X)$, where (x_n) is the solution of the (5.4.1). Moreover,

$$\Delta^2 x_n = \mathcal{K}_B f(n) + f_n.$$

The following is the main result of this section.

Theorem 5.4.2. *Let X be a UMD space and let $T \in \mathcal{B}(X)$ be an analytic operator such that $1 \in \rho(T)$. Then the following assertions are equivalent:*

(i) *Equation (5.4.1) has discrete maximal l_p-regularity.*

(ii) *The set $\left\{ \dfrac{(z-1)^2}{z} \left(\dfrac{(z-1)^2}{z} - (I - T)\right)^{-1} : |z| = 1, z \neq 1 \right\}$ is R-bounded.*

Proof. (ii) $\implies$ (i). By definition, (5.3.1) has discrete maximal l_p-regularity if and only if the operator

$$\mathcal{K}_B f(n) := \sum_{k=0}^{n} (I - T) B(k) f_{n-k}$$

is bounded on $l_p(\mathbb{Z}_+; X)$ for some $p \in (1, \infty)$. Define $\mathcal{K}_B f(n)$ by 0 for $n < 0$. Taking the time-discrete Fourier transform, we obtain by (5.3.3):

$$\widehat{\mathcal{K}_B f}(z) = (I - T)\left(\frac{(z-1)^2}{z} - (I - T)\right)^{-1} \hat{f}(z).$$

From the identity

$$(I - T)\left(\frac{(z-1)^2}{z} - (I - T)\right)^{-1} = \frac{(z-1)^2}{z}\left(\frac{(z-1)^2}{z} - (I - T)\right)^{-1} - I,$$

we obtain

$$\widehat{\mathcal{K}_B f}(z) = \frac{(z-1)^2}{z}\left(\frac{(z-1)^2}{z} - (I - T)\right)^{-1} \hat{f}(z) - \hat{f}(z). \qquad (5.4.2)$$

By hypothesis, the set

$$\left\{ \frac{(z-1)^2}{z}\left(\frac{(z-1)^2}{z} - (I - T)\right)^{-1} : |z| = 1, z \neq 1 \right\}$$

is R-bounded. Define

$$M(t) := \frac{(e^{it} - 1)^2}{e^{it}}\left(\frac{(e^{it} - 1)^2}{e^{it}} - (I - T)\right)^{-1}, t \in (0, 2\pi).$$

Then

$$(e^{it} - 1)(e^{it} + 1)M'(t) = \left(2i\,e^{it}(e^{it} + 1) - i(e^{it} - 1)(e^{it} + 1)\right)\left(M(t) - (M(t))^2\right).$$

We conclude by hypothesis and the permanence properties of R-bounded sets (Proposition 2.2.5) that the sets

$$\{M(t) : t \in (0, 2\pi)\} \quad \text{and} \quad \{(e^{it} - 1)(e^{it} + 1)M'(t) : t \in (0, 2\pi)\}$$

are R-bounded.

Then, by Blunck's theorem, there is an operator $T_M \in \mathcal{B}(l_p(\mathbb{Z}; X))$ such that

$$\widehat{T_M f}(z) = \frac{(z-1)^2}{z}\left(\frac{(z-1)^2}{z} - (I-T)\right)^{-1}\hat{f}(z).$$

Hence (5.4.2) implies that

$$\widehat{\mathcal{K}_B}(z) = \widehat{T_M}(z) - I.$$

Therefore

$$\mathcal{K}_B f(n) = T_M f(n) + (\delta * f)(n)$$

and, in particular, we conclude that $\mathcal{K}_B$ is bounded on $l_p(\mathbb{Z}_+; X)$, proving (i).

(i) $\implies$ (ii). Define

$$Rf(n) := (\mathcal{K}_B f)(n) + (\delta * f)(n).$$

Then R is a bounded operator in $l_p(\mathbb{Z}; X)$ for which, in view of (5.4.2), we have

$$\widehat{Rf}(z) = \widehat{\mathcal{K}_B f}(z) + \hat{f}(z) = \frac{(z-1)^2}{z}\left(\frac{(z-1)^2}{z} - (I-T)\right)^{-1}\hat{f}(z).$$

It follows from Proposition 2.5.11 that the set

$$\left\{\frac{(z-1)^2}{z}\left(\frac{(z-1)^2}{z} - (I-T)\right)^{-1} : |z| = 1, z \neq 1\right\}$$

is R-bounded.   □

Remark 5.4.3. Note that

$$\left\{\frac{(z-1)^2}{z}\left(\frac{(z-1)^2}{z} - (I-T)\right)^{-1} : |z| = 1, z \neq 1\right\}$$

is R-bounded if and only if

$$\left\{(z-1)^2\left(\frac{(z-1)^2}{z} - (I-T)\right)^{-1} : |z| = 1, z \neq 1\right\}$$

is R-bounded.

Corollary 5.4.4. *Let H be a Hilbert space and let $T \in \mathcal{B}(H)$ be an analytic operator such that $1 \in \rho(T)$. Then the following assertions are equivalent:*

(i) *Equation (5.4.1) has discrete maximal l_p-regularity.*

(ii) $\displaystyle\sup_{|z|=1, z\neq 1} \left\| \frac{(z-1)^2}{z} \left(\frac{(z-1)^2}{z} - (I - T) \right)^{-1} \right\| < \infty.$

Example 5.4.5. Letting $H = \mathbb{C}$ and $T = \rho I$ with $0 \leq \rho < 1$ we find that the hypothesis of the preceding corollary is satisfied. We conclude that the scalar equation

$$\begin{cases} \Delta^2 x_n - (1 - \rho)x_{n+1} = f_n, \ n \in \mathbb{Z}_+, \\ x_0 = x_1 = 0, \end{cases}$$

has the property that for all $(f_n) \in l_p(\mathbb{Z}_+)$ we have $(\Delta^2 x_n) \in l_p(\mathbb{Z}_+)$. In particular, $x_n \to 0$, i.e., the solution is stable. Note that using (5.3.3) we can infer that the fundamental solution is given by

$$B(n) = \frac{1}{a - b}(a^n - b^n),$$

where a and b are the real roots of $z^2 + (\rho - 3)z - 1 = 0$. Moreover, the solution is given by

$$x_{m+1} = (B * f)_m = \sum_{j=0}^{m} \frac{1}{a - b}\left(a^{(m-j)} - b^{(m-j)}\right) f(j).$$

Notice that the results of this section also hold in a more general form without the assumption $1 \in \rho(T)$ (see [38] for details).

5.5   Regularity in Weighted l_p Spaces

Let T be a bounded linear operator on X. Our main objective of this section is to characterize the maximal regularity in weighted spaces

$$l_p^r(\mathbb{Z}_+; X) := \left\{(x_n) : (r^{-n}x_n) \in l_p(\mathbb{Z}_+; X)\right\} \ (r > 0)$$

for the following discrete second-order evolution equation:

$$\Delta^2 u_n - (I - T)u_n = f_n, \quad n \in \mathbb{Z}_+, \tag{5.5.1}$$

with initial conditions and $f \in l_p^r(\mathbb{Z}_+; X)$.

We note that given $f \in l_p^r(\mathbb{Z}_+; X)$, defining $f(n) = 0$ for $n < 0$, it follows that the discrete-time Fourier transform of f is well defined for $0 < r < 1$ because for $1/p + 1/q = 1$ we have

$$\sum_{n=0}^{\infty} |e^{-itn} f(n)| = \sum_{n=0}^{\infty} |r^n e^{-itn} r^{-n} f(n)|$$

$$\leq \left(\sum_{n=0}^{\infty} |r^n|^q \right)^{1/q} \left(\sum_{n=0}^{\infty} |r^{-n} f(n)|^p \right)^{1/p} < \infty,$$

thanks to Hölder inequality.

We notice that if we set $(\tau_r x)(n) := r^{-n} x(n)$, then the following identity holds:

$$\Delta_r^2 = r^2 \, \tau_{r-1} \circ \Delta^2 \circ \tau_r, \tag{5.5.2}$$

where Δ_r denotes the r-difference operator (see Chap. 7). It shows that well-posedness of (5.5.1) in the weighted spaces $l_p^r(\mathbb{Z}_+; X)$ is equivalent to the study of the discrete-time evolution equation

$$\begin{cases} \Delta_r^2 x_n - r^2 (I - T) x_n = f_n, \, n \in \mathbb{Z}_+, \\ \\ x_0 = x, \, \Delta x_0 = y, \end{cases} \tag{5.5.3}$$

in the usual vector-valued Lebesgue space $l_p(\mathbb{Z}_+; X)$.

Proposition 5.5.1. *Let $T \in B(X)$ be given, then the (unique) solution of (5.5.3) is given by*

$$x_{m+1} = r^{m+1} C(m + 1) x + r^m S(m + 1) y + (r^{-1} S * f)_m. \tag{5.5.4}$$

Moreover,

$$\Delta_r x_{m+1} = r^{m+2} (I - T) S(m + 1) x + r^{m+1} C(m + 1) y + (r^{\cdot} C * f)_m. \tag{5.5.5}$$

Proof. Let x_n be the solution of (5.5.3) and define

$$v_n := [x_n, \Delta_r x_n], \quad F_n := [0, f_n]$$

and the operator $R_{T,r} \in B(X \times X)$ by

$$R_{T,r}[x, y] = [rx + y, \, r^2 (x - Tx) + ry].$$

Then, we can infer that the (5.5.3) is equivalent to

$$v_{n+1} - R_{T,r} v_n = F_n, \quad v_0 = [x_0, \Delta x_0] = (x, y), \tag{5.5.6}$$

which has the solution

$$v_{m+1} = R_{T,r}^{m+1} v_0 + \sum_{n=0}^{m} R_{T,r}^n F_{m-n}. \tag{5.5.7}$$

Denote

$$R_{T,r} = \begin{bmatrix} rI & I \\ r^2(I - T) & rI \end{bmatrix}.$$

Then a calculation shows us that

$$R_{T,r}^n = \begin{bmatrix} r^n C(n) & r^{n-1} S(n) \\ r^{n+1}(I - T)S(n) & r^n C(n) \end{bmatrix}.$$

The result is now a consequence of formula (5.5.7). The uniqueness follows from induction and this completes the proof. $\qquad\square$

For further use, we consider the (5.5.3) with zero initial condition, i.e.,

$$\begin{cases} \Delta_r^2 x_n - r^2(I - T)x_n = f_n, \ n \in \mathbb{Z}_+, \\ \\ x_0 = 0, \ x_1 = 0, \end{cases} \tag{5.5.8}$$

Corollary 5.5.2. *Let $T \in \mathcal{B}(X)$ be given, then the (unique) solution of (5.5.8) is given by*

$$x_{m+1} = (r^{\cdot-1}S * f)_m. \tag{5.5.9}$$

Moreover,

$$\Delta_r x_{m+1} = (r^{\cdot} C * f)_m. \tag{5.5.10}$$

Definition 5.5.3. Let $1 < p < +\infty$. We say that (5.5.8) has discrete maximal l_p-regularity if $\mathcal{K}^r f := (I - T)r^{\cdot+1}S * f$ defines a linear bounded operator $\mathcal{K}^r \in \mathcal{B}(l_p(\mathbb{Z}_+; X))$.

Note that, in particular, the above definition implies that for all $(f_n) \in l_p(\mathbb{Z}_+; X)$ we have $(\Delta_r^2 x_n) \in l_p(\mathbb{Z}_+; X)$ where x_n is the solution of (5.5.8).

Proposition 5.5.4. *Let $T \in \mathcal{B}(X)$ be an analytic operator. Then*

$$\sigma(r^2(I - T)) \subseteq \mathbb{D}(r^2, r^2) \cup \{0\}.$$

In particular $(z - r)^2 \in \rho(r^2(I - T))$ whenever $|z| = \alpha r$, $\alpha = 1 + \sqrt{2}$, $z \neq \alpha r$.

Proof. Let $z \notin \mathbb{D}(r^2, r^2) \cup \{0\}$, then

$$\frac{z}{r^2} \notin \mathbb{D}(1, 1) \cup \{0\}.$$

By [53, Lemma 2.10], we get $\frac{z}{r^2} \in \rho(I - T)$, that is, $z \in \rho(r^2(I - T))$. Hence

$$\sigma(r^2(I - T)) \subseteq \mathbb{D}(r^2, r^2) \cup \{0\}.$$

For the last assertion, we note that

$$|(z - r)^2 - r^2| = |z||z - 2r| \geq \alpha(\alpha - 2)r^2 = (\sqrt{2} + 1)(\sqrt{2} - 1)r^2 = r^2.$$

$\square$

In what follows, we will always assume that

$$\alpha = 1 + \sqrt{2}, \ r \geq r_0, \ 1/(1 + \sqrt{2}) < r_0 < 1. \tag{5.5.11}$$

We recall that the Z-transform on $\mathbb{T}_r^\alpha := \{z \in \mathbb{C} : |z| = \alpha r\}$ is defined as

$$\mathcal{F}f(z) = \hat{f}(z) = \sum_{j=0}^{\infty} z^{-j} f(j), \quad z \in \mathbb{T}_r^\alpha.$$

We can relate the Fourier transform of f on $\mathbb{T}_r^\alpha$ with the Fourier transform of $(\alpha r)^- f$ on $\mathbb{T}_1^1$ by the formula

$$\mathcal{F}[f](\alpha r e^{it}) = \mathcal{F}[(\alpha r)^- f](e^{it}).$$

The preceding proposition enables us to prove the following properties of the Z-transform of the solution of (5.5.8).

Proposition 5.5.5. *Let* $T \in \mathcal{B}(X)$ *be an analytic operator. Then*

$$\mathcal{F}[r^{-1}S](z) = zR((z - r)^2, r^2(I - T)), \quad z \in \mathbb{T}_r^\alpha \backslash \{\alpha r\}, \tag{5.5.12}$$

and

$$\mathcal{F}[r^\cdot C](z) = z(z - r)R((z - r)^2, r^2(I - T)), \quad z \in \mathbb{T}_r^\alpha \backslash \{\alpha r\}. \tag{5.5.13}$$

Proof. Given $x \in X$, we define

$$f_n = \begin{cases} x, & \text{for } n = 0, \\ \\ 0, & \text{for } n \neq 0. \end{cases}$$

We consider the following evolution problem:

$$\Delta_r^2 x_n - r^2(I - T)x_n = f_n \text{ for all } n \in \mathbb{Z}_+, \quad x_0 = x_1 = 0. \qquad (5.5.14)$$

By Corollary 5.5.2 the (unique) solution is given by

$$x_{n+1} = (r^{\cdot -1}S * f)(n).$$

Then[3]

$$z\hat{x}(z) = \mathcal{F}[r^{\cdot -1}S](z)x, \ z \in \mathbb{T}_r^\alpha.$$

On the other hand, we note that

$$\widehat{\Delta_r x}(z) = (z - r)\hat{x}(z),$$

for $z \in \mathbb{T}_r^\alpha$. Hence, applying the Z-transform in (5.5.14) we get

$$x = (z - r)^2 \hat{x}(z) - r^2(I - T)\hat{x}(z).$$

and then multiplying by z, we obtain

$$zx = ((z - r)^2 - r^2(I - T))\mathcal{F}[r^{\cdot -1}S](z)x, \ z \in \mathbb{T}_r^\alpha,$$

whence

$$\mathcal{F}[r^{\cdot -1}S](z)x = z((z - r)^2 - r^2(I - T))^{-1}x,$$

obtaining the first assertion. To prove the second one, we note that by Corollary 5.5.2

$$\Delta_r x_{n+1} = (r^{\cdot}C * f)(n)$$

and then

$$z(z - r)\hat{x}(z) = \mathcal{F}[r^{\cdot}C](z)x, \ z \in \mathbb{T}_r^\alpha.$$

Therefore, applying Z-transform in (5.5.14) and then multiplying the result by $z(z - r)$, we get the second assertion and the proof is finished. $\square$

Next, we define the following sequence spaces; for $r \geq r_0$

$$l_{p,r}^1(\mathbb{Z}_+; X) := \Big\{ y = (y_n) : y_0 = 0, (\Delta_r y_n) \in l_p(\mathbb{Z}_+; X) \Big\},$$

---

[3]A direct calculation shows that $\hat{f}(z) = x$

$$l^2_{p,r}(\mathbb{Z}_+; X) := \left\{ y = (y_n) : y_0 = y_1 = 0, (\Delta^2_r y_n) \in l_p(\mathbb{Z}_+; X) \right\},$$

$$l_{p,I-T}(\mathbb{Z}_+; X) := \left\{ y = (y_n) : ((I - T)y_n) \in l_p(\mathbb{Z}_+; X) \right\}.$$

To state the next result, we need to introduce some notations:

$$\Theta(r) = \begin{cases} \dfrac{2-r}{(1-r)^{2-\frac{1}{p}}}, & \text{for } r < 1, \\[3mm] \dfrac{r}{(r-1)^{2-\frac{1}{p}}}, & \text{for } r > 1, \\[3mm] 1, & \text{for } r = 1, \end{cases} \qquad (5.5.15)$$

$$\gamma(r, n) = \begin{cases} 1, & \text{for } r < 1, \\[2mm] nr^n, & \text{for } r > 1, \\[2mm] n^2, & \text{for } r = 1. \end{cases} \qquad (5.5.16)$$

Proposition 5.5.6. *We have the following properties:*

(i) *For each $y \in l^2_{p,r}(\mathbb{Z}_+; X)$ and $n \in \mathbb{Z}_+$, we have the following a priori estimate:*

$$\|y_n\|_X + \|\Delta_r y_n\|_X \le \Theta(r)\gamma(r, n)\|\Delta^2_r y.\|_p. \qquad (5.5.17)$$

(ii) *Assume that (5.5.11) is fulfilled. For each $y \in l^i_{p,r}(\mathbb{Z}_+; X)$, $i = 1, 2$, the Z-transform of y is well defined in $\mathbb{T}^\alpha_r$.*

Proof. We only outline the arguments (see [37, 39] for details).

(a) We can see that (5.5.17) follows from the following estimates:

$$\|y_n\|_X \le \begin{cases} \dfrac{1}{(1-r)^{2-\frac{1}{p}}}\|\Delta^2_r y.\|_p, & \text{for } r < 1, \\[3mm] \dfrac{1}{(r-1)^{2-\frac{1}{p}}}(n-1)r^n\|\Delta^2_r y.\|_p, & \text{for } r > 1, \\[3mm] \dfrac{n(n-1)}{2}\|\Delta^2 y.\|_p, & \text{for } r = 1, n \ge 2, \end{cases} \qquad (5.5.18)$$

and

$$
||\Delta_r y_n||_X \leq
\begin{cases}
\dfrac{1}{(1-r)^{1-\frac{1}{p}}}||\Delta_r^2 y.||_p, & \text{for } r < 1, \\[3mm]
\dfrac{r^n}{(r-1)^{1-\frac{1}{p}}}||\Delta_r^2 y.||_p, & \text{for } r > 1, \\[3mm]
n||\Delta^2 y.||_p, & \text{for } r = 1, \ n \geq 1.
\end{cases}
\tag{5.5.19}
$$

(b) For $y \in l_{p,r}^1(\mathbb{Z}_+; X)$, we have

$$
||\hat{y}(z)||_X \leq \sum_{j=1}^{\infty} (\frac{1}{\alpha r})^j \sum_{i=0}^{j-1} r^i ||\Delta_r y_{j-1-i}||_X
\tag{5.5.20}
$$

$$
\leq C(\alpha, r)||\Delta_r y.||_\infty,
$$

where $C(\alpha, r)$ is a constant depending on α and r. On the other hand, for $y \in l_{p,r}^2(\mathbb{Z}_+; X)$, we have

$$
||\hat{y}(z)||_X \leq \sum_{j=2}^{\infty} (\frac{1}{\alpha r})^j \sum_{i=0}^{j-2} r^i \sum_{k=0}^{j-i-2} r^k ||\Delta_r^2 y.||_\infty
\tag{5.5.21}
$$

$$
\leq \tilde{C}(\alpha, r)||\Delta_r^2 y.||_\infty,
$$

where $\tilde{C}(\alpha, r)$ is a constant depending on α and r and the proof of (ii) is finished.
$\square$

In the following definition we denote $l_{p,r}^0(\mathbb{Z}_+; X) := l_p(\mathbb{Z}_+; X)$.

Definition 5.5.7. Assume that (5.5.11) is fulfilled. We say that $\{Q(z)\}_{z \in \mathbb{T}_r^\alpha}$ is a $l_p - l_{p,r}^i$–multiplier, $i = 0, 1, 2$, if for each $f = (f_n) \in l_p(\mathbb{Z}_+; X)$ there is a sequence $y = (y_n) \in l_{p,r}^i(\mathbb{Z}_+; X)$ such that $\hat{y}(z) = Q(z)\hat{f}(z)$, $z \in \mathbb{T}_r^\alpha$.

We have the following proposition.

Proposition 5.5.8. *Assume that (5.5.11) is fulfilled. Then the following assertions are equivalent:*

(i) $\{Q(z)\}_{z \in \mathbb{T}_r^\alpha}$ *is a* $l_p - l_{p,r}^i$*–multiplier,* $i = 1, 2$.
(ii) $\{(z-r)^i Q(z)\}_{z \in \mathbb{T}_r^\alpha}$ *is a* $l_p - l_p$*–multiplier,* $i = 1, 2$.

Proof. We only consider the case $i = 1$ (the case $i = 2$ we left to the reader). To prove (i) $\Rightarrow$ (ii), for $f = (f_n) \in l_p(\mathbb{Z}_+; X)$, there is a sequence $y = (y_n) \in l_{p,r}^1(\mathbb{Z}_+; X)$ such that

$$\hat{y}(z) = Q(z)\hat{f}(z), \; z \in \mathbb{T}_r^{\alpha}.$$

Putting $x_0 = 0$, and $x_n = \Delta_r y_n$, $n \geq 1$, we have $x = (x_n) \in l_p(\mathbb{Z}_+; X)$ and

$$\hat{x}(z) = \widehat{\Delta_r y}(z) = (z - r)Q(z)\hat{f}(z).$$

To prove $(ii) \Rightarrow (i)$, for $f = (f_n) \in l_p(\mathbb{Z}_+; X)$, there is a sequence $y = (y_n) \in l_p(\mathbb{Z}_+; X)$ such that

$$\hat{y}(z) = (z - r)Q(z)\hat{f}(z).$$

Let $x = (x_n) \in l^1_{p,r}(\mathbb{Z}_+; X)$ be a sequence such that

$$\begin{cases} \Delta_r x_n = y_n, \\[2mm] x_0 = 0. \end{cases}$$

Then,

$$(z - r)\hat{x}(z) = \widehat{\Delta_r x}(z) = (z - r)Q(z)\hat{f}(z).$$

Hence, $Q(z)$ is a $l_p - l^1_{p,r}$–multiplier. This completes the proof of the proposition.
$\qquad\qquad\qquad\qquad\qquad\qquad\qquad\qquad\qquad\qquad\qquad\qquad\qquad\qquad\quad$ □

5.6   Well-Posedness and the Maximal Regularity Space

The following is a natural extension of the concept of well-posedness from the continuous to the discrete case.

Definition 5.6.1. We say that problem (5.5.8) is well posed if for each $f = (f_n) \in l_p(\mathbb{Z}_+; X)$ there is a unique solution $x = (x_n) \in l^2_{p,r}(\mathbb{Z}_+; X) \cap l_{p,I-T}(\mathbb{Z}_+; X)$ of (5.5.8).

We observe that the space

$$MR_p(\mathbb{Z}_+; X) := l^2_{p,r}(\mathbb{Z}_+; X) \cap l_{p,I-T}(\mathbb{Z}_+; X) \qquad (5.6.1)$$

becomes a Banach space under the norm

$$\|x\|_{MR_p} := \|\Delta_r^2 x\|_p + \|r^2(I - T)x\|_p, \qquad (5.6.2)$$

and such space is called space of maximal regularity.

Proposition 5.6.2. *Let X be a Banach space and let $T \in \mathcal{B}(X)$ be an analytic operator; assume that (5.5.11) is fulfilled and suppose that problem (5.5.8) is well posed. Then*

(a) $(z - r)^2 \in \rho(r^2(I - T))$ *whenever* $|z| = \alpha r,\ z \neq \alpha r.$
(b) *The set*

$$\left\{ M(z) := (z - r)^2\left((z - r)^2 - r^2(I - T)\right)^{-1} : |z| = r\alpha,\ z \neq \alpha r \right\}$$

is R-bounded.

The following is the main result of this section. It shows that the converse of the above proposition is valid in *UMD* spaces.

Theorem 5.6.3. *Let X be a UMD space and let $T \in \mathcal{B}(X)$ be an analytic operator; assume that (5.5.11) is fulfilled. Then, the following assertions are equivalent:*

(i) *Equation (5.5.8) is well posed.*
(ii) $\mathcal{M} := \left\{ M(z) : |z| = \alpha r,\ z \neq \alpha r \right\}$ *is a $l_p - l_p$–multiplier.*
(iii) *The set $\mathcal{M}$ is R-bounded.*
(iv) *Equation (5.5.8) has discrete maximal l_p-regularity.*

We note that the equivalence (i) and (ii) in Theorem 5.6.3 is valid without the hypothesis of *UMD* space (see [39]).

Since Hilbert spaces are *UMD* spaces, we obtain as immediate consequence the following corollary.

Corollary 5.6.4. *Let H be a Hilbert space and let $T \in \mathcal{B}(H)$ be an analytic operator; assume that (5.5.11) is fulfilled. Then, the following assertions are equivalent:*

(i) *Equation (5.5.8) is well posed.*
(ii) $\displaystyle \sup_{z \in T_r^\alpha,\, z \neq \alpha.r} \left\| (z - r)^2\left((z - r)^2 - r^2(I - T)\right)^{-1} \right\| < \infty.$
(iii) *Equation (5.5.8) has discrete maximal l_p-regularity.*

Corollary 5.6.5. *Under the conditions of Theorem 5.6.3, if the problem (5.5.8) is well posed, then for each $x \in X$ and $n \in \mathbb{Z}_+$, we have the following a priori estimate:*

$$||r^{n-1}S(n)x||_X + ||r^n C(n)x||_X \leq (1 + ||\mathcal{K}^r||_{\mathcal{B}(l_p(\mathbb{Z}_+;X))})\Theta(r)\gamma(r, n + 1)||x||_X, \tag{5.6.3}$$

where $\mathcal{K}^r$ is given in Definition 5.5.3, $\Theta(r)$ and $\gamma(r,n)$ are given by (5.5.15) and (5.5.16) respectively.

Proof. Given $x \in X$, we define

$$f_n = \begin{cases} x & \text{for } n = 0, \\ 0 & \text{for } n \neq 0. \end{cases}$$

We consider the problem

$$\Delta_r^2 x_n - r^2(I - T)x_n = f_n \text{ for all } n \in \mathbb{Z}_+, \quad x_0 = x_1 = 0. \tag{5.6.4}$$

By Corollary 5.5.2, the (unique) solution $x = (x_n)$ of (5.6.4) is given by

$$x_{n+1} = r^{n-1} \mathcal{S}(n)x \text{ and } \Delta x_{n+1} = r^n \mathcal{C}(n)x.$$

Since (5.5.8) is well posed, the solution is in $l_{p,r}^2(\mathbb{Z}_+; X) \cap l_{p,I-T}(\mathbb{Z}_+; X)$; using the Proposition 5.5.6 (i), we find

$$||r^{n-1}\mathcal{S}(n)x||_X + ||r^n\mathcal{C}(n)x||_X \leq \Theta(r)\gamma(r, n + 1)||\Delta_r^2 x.||_p$$
$$\leq (1 + ||\mathcal{K}^r||_{\mathcal{B}(l_p(\mathbb{Z}_+;X))})\Theta(r)\gamma(r, n + 1)||x||_X,$$

and now the proof is finished[4]. $\square$

5.7 Comments

The sequence $\{B(n)\}_n \subseteq \mathcal{B}(X)$ defined in Sect. 5.3 is called the fundamental solution of (5.4.1) (see [38]) and satisfies the following equation:

$$\begin{cases} \Delta^2 B(n) - (I - T)B(n + 1) = 0, \\ B(0) = 0, B(1) = I. \end{cases} \tag{5.7.1}$$

We emphasize that from a more theoretical perspective a similar analysis as in Sect. 5.4 can be carried out when we consider more general initial conditions, but the price to pay for this is that the proof would certainly require additional l_p-summability condition on $\{B(n)\}_n$[5].

---

[4]

$$||\Delta_r^2 x.||_p = \left(\sum_{n=0}^{\infty} ||f_n + r^2(I-T)x_n||_X^p\right)^{1/p} \leq ||x|| + ||\mathcal{K}^r f||_p \leq (1 + ||\mathcal{K}^r||_{\mathcal{B}(l_p(\mathbb{Z}_+;X))})||x||.$$

[5] Let $V_n := [x_n, \Delta x_n]$, $F_n = [0, f_n]$ and $R_T \in \mathcal{B}(X \times X)$ defined by $R_T[x, y] = [x + y, x + 2y - T(x + y)]$. Equation (5.4.1) is equivalent to

It should be interesting to study algebraic properties of the sequence $\{B(n)\}_n$ analogous to the case $\{C(n)\}_n$. In particular, is there a corresponding functional equation associated to $\{B(n)\}_n$?

Proposition 5.5.5 in Sect. 5.5 is from [39]. The proof of Proposition 5.6.2 in Sect. 5.6 is given in [39].

We note that when (5.5.8) has discrete maximal regularity, then $(x_n) \in MR_p(\mathbb{Z}_+; X)$ whenever $(f_n) \in l_p(\mathbb{Z}_+; X)$. It establishes an isomorphism between the set of data (f_n) and the set of solutions (x_n).

For each $f = (f_n) \in l_p(\mathbb{Z}_+; X)$ and under the equivalent conditions of Theorem 5.6.3, there exists a constant $C > 0$ such that

$$\|\Delta_r^2 x\|_p + \|r^2(I - T)x\|_p \leq C\|f\|_p,$$

for all $x = (x_n) \in MR_p(\mathbb{Z}_+; X)$, that is, the application

$$\mathcal{R} : l_p(\mathbb{Z}_+; X) \to MR_p(\mathbb{Z}_+; X)$$

given by $(f_n) \mapsto (x_n)$ is continuous.

Characterizations of maximal regularity for linear difference equations of higher order and for linear Volterra difference equations are not yet developed and it should be interesting to have a corresponding theory in these cases (see Sect. 7.7). The case of fractional linear difference equations is still an open problem, where even the concept of fractional difference equations is still in their infancy.

Maximal Regularity of Second-Order Delay Equations

Bu and Fang in [27] have studied the maximal regularity problem for second-order delay equations in Banach spaces. They gave necessary and sufficient conditions for maximal L^p-regularity (resp., Besov regularity or Triebel–Lizorkin regularity) for the following inhomogeneous abstract delay equation:

$$\begin{cases} V_{n+1} - R_T V_n = F_n, n \in \mathbb{Z}_+, \\ \quad\quad V_0 = [0, 0], \end{cases}$$

which has the solution $V_{m+1} = \sum_{n=0}^{m} R_T^n F_{m-n}$. We can see that there is an operator $B(n) \in \mathcal{B}(X)$ with $(I - T)B(n) = B(n)(I - T)$ such that

$$R_T^n = \begin{bmatrix} \Delta B(n) - B(n)(I - T) & B(n) \\ B(n)(I - T) & \Delta B(n) \end{bmatrix}.$$

where $B(n)$ satisfy the (5.7.1). On the other hand, we have that $x_{m+1} = (B * f)_m$, and $\Delta x_{m+1} = (\Delta B * f)_m$.

$$(\mathcal{P}) \quad \begin{cases} u''(t) = Au(t) + Gu'_t + Fu_t + f(t), \ t \in [0, 2\pi], \\ u(0) = u(2\pi), \ u'(0) = u'(2\pi), \end{cases} \tag{5.7.2}$$

where A is a closed linear operator in a Banach space X, $f \in \mathcal{F}([0, 2\pi]; X)$, where $\mathcal{F}([0, 2\pi]; X)$ is one of the spaces $L^p([0, 2\pi]; X)$, $B^s_{p,q}([0, 2\pi]; X)$, or $F^s_{p,q}([0, 2\pi]; X)$, and F and G are delay X-valued bounded linear operators on $\mathcal{F}([-2\pi, 0]; X)$. Moreover, for $t \in [0, 2\pi]$, u_t is an element of $\mathcal{F}([-2\pi, 0]; X)$ defined by

$$u_t(s) = u(t + s) \text{ for } -2\pi \le s \le 0,$$

here we identify a function u on $[0, 2\pi]$ with its natural 2π-periodic extension on $\mathbb{R}$. For u'_t, we know that $(u_t)'(s) = (u')_t(s)$ for $0 \le t \le 2\pi$ and $-2\pi \le s \le 0$.

Bu in [29] has studied maximal C^α-regularity $(0 < \alpha < 1)$ for the problem $(\mathcal{P})$ on $\mathbb{R}$, that is, using known C^α-multiplier result, he gave necessary and sufficient conditions for the second-order delay equation:

$$u''(t) = Au(t) + Gu'_t + Fu_t + f(t), \ t \in \mathbb{R}, \tag{5.7.3}$$

to have maximal regularity [6] in Hölder continuous function spaces $C^\alpha(\mathbb{R}; X)$, where A is a closed operator in a Banach space X, $f \in C^\alpha(\mathbb{R}; X)$ is given, and for some fixed $r > 0$, $F, G \in \mathcal{B}(C([-r, 0], X), X)$ are delay operators. Moreover $u_t, u'_t \in C([-r, 0], X)$ are given by $u_t(s) = u(t + s)$ and $u'_t(s) = u'(t + s)$ for $-r \le s \le 0$ and $t \in \mathbb{R}$, $C([-r, 0], X)$ is the space of all X-valued continuous functions on $[-r, 0]$ equipped with the norm $\|v\|_\infty := \sup_{-r \le t \le 0} \|v(t)\|$ so that it becomes a Banach space.

Recently, Lizama [134] obtained necessary and sufficient conditions for the first-order delay equation $u'(t) = Au(t) + Fu_t + f(t)$, $t \in [0, 2\pi]$ to have maximal L^p-regularity using Marcinkiewicz operator-valued multiplier theorem (Theorem 2.4.8) on $L^p([0, 2\pi]; X)$. Maximal C^α-regularity of the corresponding equation on the real line has been studied by Lizama and Poblete [135]. We note that in the special case when $G = F = 0$, maximal L^p-regularity and maximal $B^s_{p,q}$-regularity of $(\mathcal{P})$ have been studied by Arendt and Bu [8, 9] and maximal $F^s_{p,q}$-regularity of $(\mathcal{P})$ have been studied by Bu and Kim [28]. Maximal regularity of second-order equations with periodic condition has been also studied by Keyantuo and Lizama [119].

Definition 5.7.1. Let $1 \le p < \infty$ and $f \in L^p([0, 2\pi]; X)$ be given. A function $u \in H^{2,p}([0, 2\pi]; X)$ is called a strong L^p-solution of $(\mathcal{P})$, if $u(t) \in D(A)$ and $(\mathcal{P})$ holds $a.e.$ on $[0, 2\pi]$, $Au \in L^p([0, 2\pi]; X)$, and the functions $t \to Fu_t, t \to Gu'_t$

---

[6]The problem (5.7.3) is said to have maximal C^α-regularity if for each $f \in C^\alpha(\mathbb{R}; X)$ there is a unique $u \in C^{2+\alpha}(\mathbb{R}; X) \cap C^\alpha(\mathbb{R}; D(A))$ such that (5.7.3) is satisfied, and $Fu, Gu' \in C^\alpha(\mathbb{R}; X)$.

also belong to $L^p([0, 2\pi]; X)$. We say that $(\mathcal{P})$ has maximal L^p-regularity if for each $f \in L^p([0, 2\pi]; X)$ $(\mathcal{P})$ has a unique strong L^p-solution.

Let $F, G \in \mathcal{B}(L^p([-2\pi, 0]; X), X)$ and $k \in \mathbb{Z}$. We define the operators $F_k, G_k \in \mathcal{B}(X)$ by $F_k x := F(e_k \otimes x)$, $G_k x := G(e_k \otimes x)$, for all $x \in X$. It is clear that

$$\|F_k\| \le (2\pi)^{1/p}\|F\| \text{ and } \|G_k\| \le (2\pi)^{1/p}\|G\|.$$

By [134], the set $\{F_k : k \in \mathbb{Z}\}$ is R-bounded. We define the spectrum of $(\mathcal{P})$ by

$$Spec(\mathcal{P}) := \{k \in \mathbb{Z} : -k^2 I - ikG_k - F_k - A \text{ is not invertible from } D(A) \text{ into } X\}.$$

Since A is closed, if $k \in \mathbb{Z} \backslash Spec(\mathcal{P})$, then $(-k^2 I - ikG_k - F_k - A)^{-1}$ is a bounded linear operator on X. This is an easy consequence of the closed graph theorem.

We will use the following notations: for $k \in \mathbb{Z} \backslash Spec(\mathcal{P})$,

$$N_k := (-k^2 I - ikG_k - F_k - A)^{-1}, \quad M_k := -k^2 N_k. \tag{5.7.4}$$

The following necessary condition for $(\mathcal{P})$ to have maximal L^p-regularity has been obtained in [27, Proposition 2.6].

Proposition 5.7.2. *Let A be a closed linear operator defined in a Banach space X, $1 \le p < \infty$, $F, G \in \mathcal{B}(L^p([-2\pi, 0]; X), X)$. Assume that the problem $(\mathcal{P})$ has maximal L^p-regularity. Then $Spec(\mathcal{P}) = \emptyset$ and the set $\{M_k : k \in \mathbb{Z}\}$ is R-bounded.*

The following result is due to Bu and Fang [27], which completely characterizes maximal L^p-regularity of $(\mathcal{P})$.

Theorem 5.7.3. *Let A be a closed linear operator defined in a UMD Banach space X, $1 < p < \infty$, $F, G \in \mathcal{B}(L^p([-2\pi, 0]; X), X)$. Then the following assertions are equivalent:*

(a) *$(\mathcal{P})$ has maximal L^p-regularity.*
(b) *$Spec(\mathcal{P}) = \emptyset$ and the set $\{M_k : k \in \mathbb{Z}\}$ is R-bounded.*

Before introducing maximal $B_{p,q}^s$-regularity for $(\mathcal{P})$, we recall that if $f \in B_{p,q}^s([0, 2\pi]; X)$, we identify f with its periodic extension to $\mathbb{R}$. In this way, if $r \in [0, 2\pi]$ is fixed, we say that a function $f : [r, r + 2\pi] \to X$ is in $B_{p,q}^s([r, r + 2\pi]; X)$ if and only if its periodic extension to $\mathbb{R}$ is in $B_{p,q}^s([0, 2\pi]; X)$. It is easy to verify from the definition that if $f \in B_{p,q}^s([0, 2\pi]; X)$ and $t_0 \in [0, 2\pi]$ is fixed, then the function f_{t_0} defined on $[-2\pi, 0]$ by $f_{t_0}(t) = f(t_0 + t)$ is still an element of $B_{p,q}^s([0, 2\pi]; X)$ and $\|f_{t_0}\|_{B_{p,q}^s} = \|f\|_{B_{p,q}^s}$.

Let $1 \le p, q \le \infty$, $s > 0$ be fixed. We consider the problem $(\mathcal{P})$ where $f \in B_{p,q}^s([0, 2\pi]; X)$ is given, and $F, G : B_{p,q}^s([-2\pi, 0]; X) \to X$ are bounded linear operators. Moreover, for fixed $t \in [0, 2\pi]$, u_t is an element of $B_{p,q}^s([-2\pi, 0]; X)$ defined by $u_t(h) = u(t + h)$ for $-2\pi \le h \le 0$.

Definition 5.7.4. Let $1 \leq p, q \leq \infty$, $s > 0$ and $f \in B_{p,q}^s([0, 2\pi]; X)$ be given. A function $u \in B_{p,q}^{s+2}([0, 2\pi]; X)$ is called a strong $B_{p,q}^s$-solution of $(\mathcal{P})$, if $u(t) \in D(A)$ and $(\mathcal{P})$ holds *a.e.* on $[0, 2\pi]$, $Au \in B_{p,q}^s([0, 2\pi]; X)$, and the functions $t \to Fu_t$, $t \to Gu_t'$ also belong to $B_{p,q}^s([0, 2\pi]; X)$. We say that $(\mathcal{P})$ has maximal $B_{p,q}^s$-regularity if for each $f \in B_{p,q}^s([0, 2\pi]; X)$ $(\mathcal{P})$ has a unique strong $B_{p,q}^s$-solution.

Remark 5.7.5. From Remark 2.5.23 (a), when $s > 0$, we have

$$B_{p,q}^s([0, 2\pi]; X) \subset L^p([0, 2\pi]; X)$$

and the inclusion is continuous. Thus when $u \in B_{p,q}^{s+2}([0, 2\pi]; X)$ is a strong $B_{p,q}^s$-solution of $(\mathcal{P})$, then $u \in H^{2,p}([0, 2\pi]; X)$; therefore u is twice differentiable *a.e.* and $u(0) = u(2\pi)$, $u'(0) = u'(2\pi)$.

Let $F, G \in \mathcal{B}(B_{p,q}^s([-2\pi, 0]; X), X)$ and $k \in \mathbb{Z}$. We define the operators $F_k, G_k \in \mathcal{B}(X)$ by $F_k x := F(e_k \otimes x)$, $G_k x := G(e_k \otimes x)$, for all $x \in X$, $k \in \mathbb{Z}$. It is clear that there is a constant $C > 0$ such that $\|e_k \otimes x\|_{B_{p,q}^s} \leq C\|x\|$ for all $k \in \mathbb{Z}$. Thus

$$\|F_k\| \leq C\|F\| \text{ and } \|G_k\| \leq C\|G\|.$$

We will use the following notation: for $k \in \mathbb{Z}\backslash Spec(\mathcal{P})$, let N_k, M_k be given by (5.7.4).

The following result is due to Bu and Fang [27], which characterizes maximal $B_{p,q}^s$-regularity of $(\mathcal{P})$.

Theorem 5.7.6. *Let A be a closed linear operator defined in a Banach space X, $1 \leq p, q \leq \infty$, $s > 0$ and $F, G \in \mathcal{B}(B_{p,q}^s([-2\pi, 0]; X), X)$. Assume that*

$$\sup_{k \in \mathbb{Z}} \|k(G_{k+2} - 2G_{k+1} + G_k)\| < \infty.$$

Then the following assertions are equivalent:

(a) *$(\mathcal{P})$ has maximal $B_{p,q}^s$-regularity.*
(b) *$Spec(\mathcal{P}) = \emptyset$ and the set $\{M_k : k \in \mathbb{Z}\}$ is uniformly bounded.*

The main ingredient to prove Theorem 5.7.6 is Theorem 2.5.29. When the underlying Banach space X is B-convex, the Marcinkiewicz estimate of order 1 is sufficient for the sequence $(M_k)_{k \in \mathbb{Z}}$ to be a $B_{p,q}^s$-multiplier. Hence we have the following result on maximal $B_{p,q}^s$-regularity of the problem $(\mathcal{P})$ when X in B-convex.

Corollary 5.7.7. *Let A be a closed linear operator defined in a B-convex Banach space X, $1 \leq p, q \leq \infty$, $s > 0$ and $F, G \in \mathcal{B}(B_{p,q}^s([-2\pi, 0]; X), X)$. Then the following assertions are equivalent:*

(a) $(\mathcal{P})$ *has maximal* $B_{p,q}^s$-*regularity.*
(b) $Spec(\mathcal{P}) = \emptyset$ *and the set* $\{M_k : k \in \mathbb{Z}\}$ *is uniformly bounded.*

Before introducing maximal $F_{p,q}^s$-regularity for $(\mathcal{P})$, we recall that if $f \in F_{p,q}^s([0, 2\pi]; X)$, we identify f with its periodic extension to $\mathbb{R}$. In this way, if $r \in [0, 2\pi]$ is fixed, we say that a function $f : [r, r + 2\pi] \to X$ is in $F_{p,q}^s([r, r + 2\pi]; X)$ if and only if its periodic extension to $\mathbb{R}$ is in $F_{p,q}^s([0, 2\pi]; X)$. It is easy to verify that if $f \in F_{p,q}^s([0, 2\pi]; X)$ and $t_0 \in [0, 2\pi]$ is fixed, then the function f_{t_0} is an element of $F_{p,q}^s([0, 2\pi]; X)$, and $\|f_{t_0}\|_{F_{p,q}^s} = \|f\|_{F_{p,q}^s}$.

Let $1 \le p < \infty$, $1 \le q \le \infty$, $s > 0$ be fixed. We consider the problem $(\mathcal{P})$ where $f \in F_{p,q}^s([0, 2\pi]; X)$ is given, and $F, G : F_{p,q}^s([-2\pi, 0]; X) \to X$ are fixed bounded linear operators. Moreover, for fixed $t \in [0, 2\pi]$, u_t is an element of $F_{p,q}^s([-2\pi, 0]; X)$.

Now we give the definitions of strong $F_{p,q}^s$-solutions and maximal $F_{p,q}^s$-regularity of $(\mathcal{P})$ (see [27]).

Definition 5.7.8. Let $1 \le p < \infty$, $1 \le q \le \infty$, $s > 0$ and $f \in F_{p,q}^s([0, 2\pi]; X)$ be given. A function $u \in F_{p,q}^{s+2}([0, 2\pi]; X)$ is called a strong $F_{p,q}^s$-solution of $(\mathcal{P})$ if $u(t) \in D(A)$ and $(\mathcal{P})$ holds *a.e.* on $[0, 2\pi]$, $Au \in F_{p,q}^s([0, 2\pi]; X)$, and the functions $t \to Fu_t$, $t \to Gu'_t$ also belong to $F_{p,q}^s([0, 2\pi]; X)$. We say that $(\mathcal{P})$ has maximal $F_{p,q}^s$-regularity if for each $f \in F_{p,q}^s([0, 2\pi]; X)$ $(\mathcal{P})$ has a unique strong $F_{p,q}^s$-solution.

Remark 5.7.9. From Proposition 2.5.31 (c), when $s > 0$, we have

$$F_{p,q}^s([0, 2\pi]; X) \subset L^p([0, 2\pi]; X)$$

and the inclusion is continuous. Thus when $u \in F_{p,q}^{s+2}([0, 2\pi]; X)$ is a strong $F_{p,q}^s$-solution of $(\mathcal{P})$, then $u \in H^{2,p}([0, 2\pi]; X)$; therefore u is twice differentiable *a.e.* and $u(0) = u(2\pi)$, $u'(0) = u'(2\pi)$.

Let $F, G \in \mathcal{B}(F_{p,q}^s([-2\pi, 0]; X), X)$ and $k \in \mathbb{Z}$. We define the operators $F_k, G_k \in \mathcal{B}(X)$ by

$$F_k x := F(e_k \otimes x), \quad G_k x := G(e_k \otimes x),$$

for all $x \in X$, $k \in \mathbb{Z}$. It is clear that there is a constant $C > 0$ such that

$$\|e_k \otimes x\|_{F_{p,q}^s} \le C\|x\|$$

for all $k \in \mathbb{Z}$. Thus

$$\|F_k\| \le C\|F\| \text{ and } \|G_k\| \le C\|G\|.$$

As in the Lebesgue and Besov space cases, we will use the following notation: for $k \in \mathbb{Z} \setminus Spec(\mathcal{P})$, let N_k, M_k be given by (5.7.4).

The following result is due to Bu and Fang [27], which characterizes maximal $F_{p,q}^s$-regularity of $(\mathcal{P})$.

Theorem 5.7.10. *Let A be a closed linear operator defined in a Banach space X, $1 \leq p < \infty$, $1 \leq q \leq \infty$, $s > 0$ and $F, G \in \mathcal{B}(F_{p,q}^s([-2\pi, 0]; X), X)$. Assume that*

$$\sup_{k \in \mathbb{Z}} \|k(G_{k+2} - 2G_{k+1} + G_k)\| < \infty,$$

$$\sup_{k \in \mathbb{Z}} \|k^2(G_{k+3} - 3G_{k+2} + 3G_{k+1} - G_k)\| < \infty,$$

and

$$\sup_{k \in \mathbb{Z}} \|k(F_{k+3} - 3F_{k+2} + 3F_{k+1} - F_k)\| < \infty.$$

Then the following assertions are equivalent:

(a) $(\mathcal{P})$ *has maximal $F_{p,q}^s$-regularity.*
(b) $Spec(\mathcal{P}) = \emptyset$ *and the set $\{M_k : k \in \mathbb{Z}\}$ is uniformly bounded.*

The main ingredient to prove Theorem 5.7.10 is Theorem 2.5.33. When the underlying Banach space X is B-convex, the Marcinkiewicz estimate of order 2 is sufficient for the sequence $(M_k)_{k \in \mathbb{Z}}$ to be a $F_{p,q}^s$-multiplier. We have hence the following result on maximal $F_{p,q}^s$-regularity of the problem $(\mathcal{P})$ when X is B-convex.

Corollary 5.7.11. *Let A be a closed linear operator defined in a B-convex Banach space X, $1 \leq p < \infty$, $1 \leq q \leq \infty$, $s > 0$, and let F, G be in $\mathcal{B}(F_{p,q}^s([-2\pi, 0]; X), X)$. Assume that*

$$\sup_{k \in \mathbb{Z}} \|k(G_{k+2} - 2G_{k+1} + G_k)\| < \infty.$$

Then the following assertions are equivalent:

(a) $(\mathcal{P})$ *has maximal $F_{p,q}^s$-regularity.*
(b) $Spec(\mathcal{P}) = \emptyset$ *and the set $\{M_k : k \in \mathbb{Z}\}$ is uniformly bounded.*

Chapter 6
Second-Order Semilinear Difference Equations

In this chapter we are concerned with the study of the existence of bounded solutions for certain second-order semilinear difference equations by means of the knowledge of maximal regularity properties for the associated homogeneous discrete-time evolution equation.

The general framework for the proof of our statements uses a novel approach based on the characterizations of discrete maximal regularity obtained in the previous chapter. The implementation of this approach is not trivial because we need to obtain a priori estimates and to define suitable discrete Sobolev spaces.

6.1 Semilinear Second-Order Equations

In this section our aim is to investigate the existence of bounded solutions for semilinear second-order evolution equations, whose second discrete derivative is in ℓ_p.

We consider the following second-order evolution equation:

$$\begin{cases} \Delta^2 x_n - (I - T)x_n = f(n, x_n, \Delta x_n), & n \in \mathbb{Z}_+, \\ \\ x_0 = 0, \ x_1 = 0, \end{cases} \tag{6.1.1}$$

which is equivalent to

$$\begin{cases} x_{n+2} - 2x_{n+1} + Tx_n = f(n, x_n, \Delta x_n), & n \in \mathbb{Z}_+, \\ \\ x_0 = 0, \ x_1 = 0, \end{cases} \tag{6.1.2}$$

To establish our first result, we need to introduce the following assumption:

R.P. Agarwal et al., *Regularity of Difference Equations on Banach Spaces*, DOI 10.1007/978-3-319-06447-5_6, © Springer International Publishing Switzerland 2014

Assumption (A_2):   Suppose that the following conditions hold:

(i) The function $f : \mathbb{Z}_+ \times X \times X \longrightarrow X$ satisfies a Lipschitz condition on $X \times X$, i.e., for all $z, w \in X \times X$ and $n \in \mathbb{Z}_+$, we have

$$\|f(n, z) - f(n, w)\|_X \leq \alpha_n \|z - w\|_{X \times X},$$

where $\alpha := (\alpha_n) \in l_1(\mathbb{Z}_+)$.

(ii) $f(\cdot, 0, 0) \in l_1(\mathbb{Z}_+; X)$.

With the above notations we have the following result.

Theorem 6.1.1. *Assume that (A_2) holds. In addition suppose that (5.1.1) has discrete maximal l_p-regularity (Definition 5.1.1). Then, there is a unique bounded solution $x = (x_n)$ of (6.1.1) such that $(\Delta^2 x_n) \in l_p(\mathbb{Z}_+; X)$.*

In view of Theorem 6.1.1, we obtain the following result which is valid on UMD spaces.

Corollary 6.1.2. *Let X be a UMD space. Assume that condition (A_2) holds and suppose $T \in \mathcal{B}(X)$ is analytic and satisfy that the set*

$$\left\{(\lambda - 1)^2 R((\lambda - 1)^2, I - T) : |\lambda| = 1, \lambda \neq 1\right\}$$

is R-bounded. Then, there is a unique bounded solution $x = (x_n)$ of (6.1.1) such that $(\Delta^2 x_n) \in l_p(\mathbb{Z}_+; X)$.

Example 6.1.3.   Consider the semilinear problem

$$\begin{cases} \Delta^2 x_n - (I - T) x_n = q_n f(x_n), \ n \in \mathbb{Z}_+, \\ \\ x_0 = 0, \ x_1 = 0, \end{cases} \tag{6.1.3}$$

where f is defined and satisfies a Lipschitz condition with constant L on a Hilbert space H. In addition suppose $(q_n) \in l_1(\mathbb{Z}_+)$. Then assumption ($A_2$) is satisfied. In this case, applying the preceding result we obtain that if $T \in \mathcal{B}(H)$ is analytic and the set

$$\left\{(\lambda - 1)^2 R((\lambda - 1)^2, I - T) : |\lambda| = 1, \lambda \neq 1\right\}$$

is bounded, then there is a unique bounded solution $x = (x_n)$ of the (6.1.3) such that $(\Delta^2 x_n) \in l_p(\mathbb{Z}_+; H)$.

In particular, taking $T = I$ the identity operator, we obtain the following scalar result which complements those in Drozdowicz–Popenda [72].

Corollary 6.1.4. *Suppose that f is defined and satisfies a Lipschitz condition with constant L on a Hilbert space H. Let $(q_n) \in l_1(\mathbb{Z}_+; H)$, then the equation*

$$\Delta^2 x_n = q_n f(x_n), \tag{6.1.4}$$

has a unique bounded solution $x = (x_n)$ *such that* $(\Delta^2 x_n) \in l_p(\mathbb{Z}_+; H)$.

6.2   Exact Semilinear Second-Order Equations

In this section, our aim is to investigate the existence and uniqueness of bounded solutions for the following second-order evolution equation, whose second discrete derivative is in ℓ_p.

$$\begin{cases} \Delta^2 x_n - (I - T)x_{n+1} = f(n, x_n, \Delta x_n), \ n \in \mathbb{Z}_+, \\[2mm] x_0 = 0, \ x_1 = 0, \end{cases} \tag{6.2.1}$$

which is equivalent to

$$\begin{cases} x_{n+2} - (3I - T)x_{n+1} + x_n = f(n, x_n, \Delta x_n), \ n \in \mathbb{Z}_+, \\[2mm] x_0 = 0, \ x_1 = 0. \end{cases} \tag{6.2.2}$$

For this, we need to introduce the following condition.

Assumption (A_3):   Let $\alpha = (\alpha_n)$ be a positive sequence such that

$$\sum_{n=1}^{\infty} \alpha_n 2^{2n} < +\infty.$$

Suppose that the following conditions hold:

(i) The function $f : \mathbb{Z}_+ \times X \times X \longrightarrow X$ satisfies a Lipschitz condition on $X \times X$, that is, for all $z, w \in X \times X$ and $n \in \mathbb{Z}_+$, we have

$$\|f(n, z) - f(n, w)\|_X \le \alpha_n \|z - w\|_{X \times X}.$$

(ii) $f(\cdot, 0, 0) \in l_1(\mathbb{Z}_+; X)$.

Remark 6.2.1. Concerning the Lipschitz condition (i), for example, we can consider $f(n, z) = P(n)Q(z)$, where $P : \mathbb{Z}_+ \to \mathcal{B}(X \times X, X)$ and Q satisfying

$$\|Q(z) - Q(w)\|_{X \times X} \le C \|z - w\|_{X \times X}$$

and $\sum_{n=0}^{\infty} \|P(n)\| 2^{2n} < +\infty$. On the other hand, if $\alpha := (\alpha_n)$ is a positive sequence such that $\lim_{n \to \infty} \frac{\alpha_{n+1}}{\alpha_n} = \beta_0$ with $\beta_0 \in (0, 1/4)$, then $\sum_{n=0}^{\infty} \alpha_n 2^{2n} < +\infty$. For concrete examples of sequences $\alpha := (\alpha_n)$, we can take $\alpha_n = \beta_0^n/n, \alpha_n = \beta_0^n/n^{3/2} \ln n$.

With the above notations, we have the following result (see [52] for details).

Theorem 6.2.2. *Assume that* (A_3) *holds. In addition suppose that* $(5.4.1)$ *has discrete maximal* l_p-*regularity (Definition 5.4.1). Then, there is a unique solution* $x = (x_n)$ *of* (6.2.1) *such that* $(\Delta^2 x_n) \in l_p(\mathbb{Z}_+; X)$. *Moreover, one has the following a priori estimates for the solution:*

$$\sup_{n \in \mathbb{Z}_+} \left(\frac{1}{2^{2n}} (\|x_n\|_X + \|\Delta x_n\|_X) \right) \tag{6.2.3}$$
$$\leq (1 + \|\mathcal{K}_B\|) \| f(\cdot, 0, 0)\|_1 e^{(1 + \|\mathcal{K}_B\|) \sum_{k=0}^{\infty} \alpha_k 2^{2k}}$$

$$\|\Delta^2 x\|_p \leq (1 + \|\mathcal{K}_B\|)^2 \| f(\cdot, 0, 0)\|_1 e^{2(1 + \|\mathcal{K}_B\|) \sum_{k=0}^{\infty} \alpha_k 2^{2k}}, \quad 1 < p < +\infty, \tag{6.2.4}$$

$$\sup_{n \in \mathbb{Z}_+} \|((I - T)x_n, (I - T)\Delta x_n)\|_{X \times X} \tag{6.2.5}$$

$$\leq 3(1 + \|\mathcal{K}_B\|)^3 \| f(\cdot, 0, 0)\|_1 e^{5[1 + \|T\| + \|\mathcal{K}_B\|] \sum_{k=0}^{\infty} \alpha_k 2^{2k}},$$

where $\mathcal{K}_B$ *is the operator given in Definition 5.4.1.*

Proof. Denote by $\mathcal{H}_0^{2,p}$ the Banach space[1] of all sequences $V = (V_n)$ such that

$$V_0 = V_1 = 0 \text{ and } \Delta^2 V \in l_p(\mathbb{Z}_+; X)$$

equipped with the norm

$$\|\|V\|\| = \|\Delta^2 V\|_p.$$

Let V be a sequence in $\mathcal{H}_0^{2,p}$. We have the estimate[2]

---

[1] Let $(V^m)_m$ be a Cauchy sequence in $\mathcal{H}_0^{2,p}$, then $(\Delta^2 V^m)_m$ is a Cauchy sequence in $l_p(\mathbb{Z}_+; X)$. Hence there is $Y \in l_p(\mathbb{Z}_+; X)$ so that $\Delta^2 V^m \to Y$ in $l_p(\mathbb{Z}_+; X)$ as $m \to \infty$. If $Z = (Z_n)_n$ is a sequence such that $\Delta^2 Z = Y$, $Z_0 = Z_1 = 0$, then $V^m \to Z$ in $\mathcal{H}_0^{2,p}$ and $Z \in \mathcal{H}_0^{2,p}$.

[2] In fact, we have

$$\|(V_n, \Delta V_n)\|_{X \times X} = \left\| \sum_{i=0}^{n-1} \Delta V_i \right\|_X + \left\| \sum_{i=0}^{n-1} \Delta^2 V_i \right\|_X$$

$$\leq \sum_{i=0}^{n-1} \|\Delta V_i\|_X + \sum_{i=0}^{n-1} \|\Delta^2 V_i\|_X$$

$$\leq \sum_{i=0}^{n-1} i \|\Delta^2 V.\|_\infty + n \|\Delta^2 V.\|_\infty.$$

$$||(V_n, \Delta V_n)||_{X \times X} \le \frac{1}{2}(n^2 + n)|||V|||, \ n \in \mathbb{Z}_+. \tag{6.2.6}$$

From (6.2.6), we obtain the inequality

$$\sum_{n=0}^{\infty} \alpha_n^p ||(V_n, \Delta V_n)||_{X \times X}^p \le \left(\sum_{n=0}^{\infty} \alpha_n^p 2^{2np} \right) |||V|||. \tag{6.2.7}$$

From (A_3) and (6.2.7), we find $g := f(\cdot, V, \Delta V.) \in l_p(\mathbb{Z}_+; X)$.

Since (5.4.1) has discrete maximal l_p-regularity, the Cauchy problem

$$\begin{cases} z_{n+2} - (3I - T)z_{n+1} + z_n = g_n, \\ \\ z_0 = z_1 = 0, \end{cases} \tag{6.2.8}$$

has a unique solution (z_n) such that $(\Delta^2 z_n) \in l_p(\mathbb{Z}_+; X)$, which is given by

$$z_n = [\mathcal{J}V]_n = \begin{cases} 0, & \text{if } n = 0, 1, \\ \\ \sum_{k=1}^{n-1} B(k) f(n - 1 - k, V_{n-1-k}, \Delta V_{n-1-k}), & \text{if } n \ge 2. \end{cases} \tag{6.2.9}$$

We now show that the operator $\mathcal{J} : \mathcal{H}_0^{2,p} \longrightarrow \mathcal{H}_0^{2,p}$ has a unique fixed point. For V and W in $\mathcal{H}_0^{2,p}$, we can see that[3]

$$||\Delta^2 \mathcal{J}V - \Delta^2 \mathcal{J}W||_p$$

$$\le \left(\sum_{n=1}^{\infty} ||f(n, V_n, \Delta V_n) - f(n, W_n, \Delta W_n)||_X^p \right)^{1/p}$$

$$+ \left(\sum_{n=1}^{\infty} \left\| \sum_{k=1}^{n} (I - T)B(k)(f(n - k, V_{n-k}, \Delta V_{n-k}) \right. \right. \tag{6.2.10}$$

$$\left. \left. - f(n - k, W_{n-k}, \Delta W_{n-k})) \right\|_X^p \right)^{1/p}.$$

---

[3] Note that

$$\Delta^2 [\mathcal{J}V]_n = f(n, V_n, \Delta V_n) + \sum_{k=1}^{n} (I - T)B(k) f(n - k, V_{n-k}, \Delta V_{n-k}).$$

Since (5.4.1) has discrete maximal l_p-regularity, $\mathcal{K}_B$ is bounded on $l_p(\mathbb{Z}_+; X)$, using (A_3) and then (6.2.7), we obtain

$$|||\mathcal{J}V - \mathcal{J}W|||$$

$$\leq (1 + ||\mathcal{K}_B||)\Big(\sum_{n=0}^{\infty} \alpha_n^p ||((V - W)_n, \Delta(V - W)_n||_{X \times X}^p \Big)^{1/p} \tag{6.2.11}$$

$$\leq (1 + ||\mathcal{K}_B||)\Big(\sum_{n=0}^{\infty} \alpha_n 2^{2n} \Big)|||V - W|||.$$

We will consider the iterates of the operator $\mathcal{J}$. Initially we observe that[4]

$$||(I - T)B(n)|| \leq ||\mathcal{K}_B||, \ n \in \mathbb{Z}_+. \tag{6.2.12}$$

Taking into account (6.2.12) we can infer that

$$||B(k)|| \leq n^2(1 + ||\mathcal{K}_B||), \ 0 \leq k \leq n. \tag{6.2.13}$$

$$||\Delta B(k)|| \leq (1 + n)(1 + ||\mathcal{K}_B||), \ 0 \leq k \leq n. \tag{6.2.14}$$

Using (6.2.13) and (6.2.6), we find

$$||[\mathcal{J}V]_n - [\mathcal{J}W]_n||_X \leq n^2(1 + ||\mathcal{K}_B||)\Big(\sum_{j=0}^{n-1} \alpha_j 2^{2j} \Big)|||V - W|||. \tag{6.2.15}$$

---

[4]In fact, since $\mathcal{K} \in \mathcal{B}(l_p(\mathbb{Z}_+; X))$; for all $f \in l_p(\mathbb{Z}_+; X)$, we get

$$\Big[\sum_{n=0}^{\infty} \Big\| \sum_{k=0}^{n} (I - T)B(k)f_{n-k} \Big\|_X^p \Big]^{1/p} \leq ||\mathcal{K}_B|| \Big[\sum_{n=0}^{\infty} ||f_n||_X^p \Big]^{1/p}.$$

Given $x \in X$, we define

$$f_n = \begin{cases} x, & \text{for } n = 0, \\ 0, & \text{for } n \neq 0. \end{cases}$$

A direct calculation shows that

$$||(I - T)B(n)x||_X \leq ||\mathcal{K}_B||||x||_X, \ x \in X,$$

proving (6.2.12).

Using (6.2.14) and (6.2.6), we get

$$||\Delta[\mathcal{J}V]_n - \Delta[\mathcal{J}W]_n||_X \leq (1+n)(1+||\mathcal{K}_B||)\left(\sum_{j=0}^{n-1}\alpha_j 2^{2j}\right)|||V - W|||.$$

$$(6.2.16)$$

Using estimates (6.2.15) and (6.2.16), we obtain

$$||([\mathcal{J}V - \mathcal{J}W]_n, \Delta[\mathcal{J}V - \mathcal{J}W]_n)||_{X\times X}$$

$$\leq 2^{2n}(1+||\mathcal{K}_B||)\left(\sum_{j=0}^{n-1}\alpha_j 2^{2j}\right)|||V - W|||. \qquad (6.2.17)$$

Next, using estimates (6.2.6) and (6.2.17), we find

$$|||\mathcal{J}^2V - \mathcal{J}^2W||| \leq \frac{1}{2}\left((1+||\mathcal{K}_B||)\sum_{j=0}^{\infty}\alpha_j 2^{2j}\right)^2|||V - W|||. \qquad (6.2.18)$$

An induction argument shows to us that

$$||([\mathcal{J}^nV - \mathcal{J}^nW]_m, \Delta[\mathcal{J}^nV - \mathcal{J}^nW]_m)||_{X\times X}$$

$$\leq 2^{2m}(1+||\mathcal{K}_B||)^n\frac{1}{n!}\left(\sum_{j=0}^{m-1}\alpha_j 2^{2j}\right)^n|||V - W|||. \qquad (6.2.19)$$

Consequently, we get

$$|||\mathcal{J}^nV - \mathcal{J}^nW||| \leq \frac{1}{n!}\left((1+||\mathcal{K}_B||)\sum_{j=0}^{\infty}\alpha_j 2^{2j}\right)^n|||V - W|||. \qquad (6.2.20)$$

Since $\frac{1}{n!}\left((1+||\mathcal{K}_B||)\sum_{j=0}^{\infty}\alpha_j 2^{2j}\right)^n < 1$ for n sufficiently large, by the fixed-point iteration method, $\mathcal{J}$ has a unique fixed point $V \in \mathcal{H}_0^{2,p}$. Let V be the unique fixed point of $\mathcal{J}$, and then by (6.2.13) and assumption (A_3), we have

$$||V_n||_X \leq n^2(1+||\mathcal{K}_B||)\left(\sum_{k=0}^{n-1}\alpha_k||(V_k, \Delta V_k)||_{X\times X} + ||f(\cdot,0,0)||_1\right). \qquad (6.2.21)$$

On the other hand, from (6.2.14), we get

$$||\Delta V_n||_X \leq (1+n)(1+||\mathcal{K}_B||)\left(\sum_{k=0}^{n-1}\alpha_k||(V_k, \Delta V_k)||_{X\times X} + ||f(\cdot,0,0)||_1\right).$$

$$(6.2.22)$$

Using (6.2.21), together with the bounds (6.2.22), we obtain

$$\frac{1}{2^{2n}} \|(V_n, \Delta V_n)\|_{X \times X}$$

$$\leq (1 + \|\mathcal{K}_B\|) \| f(\cdot, 0, 0)\|_1 + (1 + \|\mathcal{K}_B\|) \sum_{k=0}^{n-1} \alpha_k 2^{2k} \frac{1}{2^{2k}} \|(V_k, \Delta V_k)\|_{X \times X}.$$

$$(6.2.23)$$

Then, as an application of the discrete Gronwall's inequality, one gets (6.2.3).

Next, proceeding analogously as in (6.2.11) we can see that

$$\|\Delta^2 V\|_p \leq (1 + \|\mathcal{K}_B\|) \sum_{n=0}^{\infty} \| f(n, V_n, \Delta V_n)\|_X$$

$$\leq (1 + \|\mathcal{K}_B\|) \Big(\sum_{n=0}^{\infty} \alpha_n 2^{2n} \big(\frac{1}{2^{2n}} \|(V_n, \Delta V_n)\|_{X \times X}\big) + \| f(\cdot, 0, 0)\|_1 \Big)$$

$$\leq (1 + \|\mathcal{K}_B\|) \Big((1 + \|\mathcal{K}_B\|) e^{(1 + \|\mathcal{K}_B\|) \sum_{k=0}^{\infty} \alpha_k 2^{2k}}$$

$$\times \sum_{n=0}^{\infty} \alpha_n 2^{2n} + 1 \Big) \| f(\cdot, 0, 0)\|_1$$

$$\leq (1 + \|\mathcal{K}_B\|)^2 \| f(\cdot, 0, 0)\|_1 e^{(1 + \|\mathcal{K}_B\|) \sum_{k=0}^{\infty} \alpha_k 2^{2k}} \Big(1 + \sum_{n=0}^{\infty} \alpha_n 2^{2n} \Big)$$

$$(6.2.24)$$

$$\leq (1 + \|\mathcal{K}_B\|)^2 \| f(\cdot, 0, 0)\|_1 e^{2(1 + \|\mathcal{K}_B\|) \sum_{k=0}^{\infty} \alpha_k 2^{2k}}.$$

This proves the estimate (6.2.4).

Finally, by (6.2.12) and using the estimate

$$\|(V_k, \Delta V_k)\|_{X \times X} \leq \|(T V_k, T \Delta V_k)\|_{X \times X} + \|((I - T) V_k, (I - T) \Delta V_k)\|_{X \times X},$$

we obtain

$$\|(I - T) V_n\|_X \leq \|\mathcal{K}_B\| \Big(\|T\| \big(\sum_{k=0}^{\infty} \alpha_k 2^{2k}\big) \|\|V\|\| + \| f(\cdot, 0, 0)\|_1 \Big)$$

$$(6.2.25)$$

$$+ \|\mathcal{K}_B\| \sum_{k=0}^{n-1} \alpha_k \|((I - T) V_k, (I - T) \Delta V_k)\|_{X \times X}.$$

Proceeding analogously as in (6.2.25), we obtain

$$\|(I - T)\Delta V_n\|_X \le 2\|\mathcal{K}_B\|\left(\|T\|(\sum_{k=0}^{\infty} \alpha_k 2^{2k})\|\|V\|\| + \|f(\cdot, 0, 0)\|_1\right)$$

$$\quad\quad\quad\quad\quad\quad (6.2.26)$$

$$+ 2\|\mathcal{K}_B\| \sum_{k=0}^{n-1} \alpha_k \|((I - T)V_k, (I - T)\Delta V_k)\|_{X \times X}.$$

From estimates (6.2.25) and (6.2.26) and the discrete Gronwall's inequality, we get

$$\|((I - T)V_n, (I - T)\Delta V_n)\|_{X \times X} \le 3\|\mathcal{K}_B\|\left(\|T\|(\sum_{k=0}^{\infty} \alpha_k 2^{2k})\|\|V\|\|\right.$$

$$\left. + \|f(\cdot, 0, 0)\|_1\right)e^{3\|\mathcal{K}_B\| \sum_{k=0}^{\infty} \alpha_k 2^{2k}}.$$

Then, using (6.2.24), we obtain (6.2.5). This ends the proof of the theorem.   □

In view of Theorem 5.4.2, we obtain the following result which is valid on *UMD* spaces.

Corollary 6.2.3. *Let X be a UMD space. Assume that* (A$_3$) *holds and suppose T is an analytic operator such that the set*

$$\left\{\frac{(z - 1)^2}{z} R\left(\frac{(z - 1)^2}{z}, I - T\right) : |z| = 1, z \ne 1\right\} \quad\quad (6.2.27)$$

is R-bounded. Then, there is a unique solution $x = (x_n)$ *of equation* (6.2.1) *such that* $(\Delta^2 x_n) \in l_p(\mathbb{Z}_+; X)$. *Moreover, the a priori estimates* (6.2.3),(6.2.4), *and* (6.2.5) *hold.*

Example 6.2.4. Consider the semilinear problem:

$$\begin{cases} x_{n+2} - (3I - T)x_{n+1} + x_n = q_n f(x_n), \, n \in \mathbb{Z}_+, \\ \\ x_0 = 0, \, x_1 = 0, \end{cases} \quad\quad (6.2.28)$$

where f is defined and satisfies a Lipschitz condition with constant L on a Hilbert space H. In addition, suppose $(2^{2n} q_n) \in l_1(\mathbb{Z}_+)$. Then assumption (A_3) is satisfied. In our case, applying the preceding result, we obtain that if $T \in \mathcal{B}(H)$ is an analytic operator such that the set (6.2.27) is bounded, then there is a unique solution $x = (x_n)$ of (6.2.28) such that $(\Delta^2 x_n) \in l_p(\mathbb{Z}_+, H)$. Moreover, one has

$$\max\left\{\sup_{n\in\mathbb{Z}_+}\left(\frac{1}{2^{2n}}(\|x_n\|_H+\|\Delta x_n\|_H)\right),\|\Delta^2 x\|_p\right\}\le C\|f(0)\|_H\|q\|_1 e^{LC\|q2^{2\cdot}\|_1},$$

$$(6.2.29)$$

$$\sup_{n\in\mathbb{Z}_+}\left(\|(I-T)x_n\|_H+\|(I-T)\Delta x_n\|_H\right)\le C\|f(0)\|_H\|q\|_1 e^{LC\|q2^{2\cdot}\|_1},$$

$$(6.2.30)$$

where C is a positive constant.

6.3   Semilinear Problems on Weighted Spaces

In this section our aim is to investigate the existence of solutions for the following second-order evolution equation, whose second discrete r–derivative is in ℓ_p :

$$\begin{cases} \Delta_r^2 x_n - r^2(I-T)x_n = G(n,x_n,\Delta_r x_n),\ n\in\mathbb{Z}_+, \\[2mm] x_0 = 0,\ x_1 = 0. \end{cases} \tag{6.3.1}$$

To state the next result, we need to introduce some notation:

$$A(r) := 2(1 + 2\|\mathcal{K}^r\|_{\mathcal{B}(l_p(\mathbb{Z}_+;X))})\Theta(r)\|\rho\cdot\gamma(r,\cdot)\|_1, \tag{6.3.2}$$

where $\Theta(r)$ is given in (5.5.15). We make the following assumption:

Assumption (A$_4$):   Suppose that the following conditions hold:

(i) The function $G : \mathbb{Z}_+ \times X \times X \longrightarrow X$ satisfies a Lipschitz condition in $X \times X$, i.e., for all $z, w \in X \times X$ and $n \in \mathbb{Z}_+$, we have

$$\|G(n,z) - G(n,w)\|_X \le \rho_n\|z-w\|_{X\times X},$$

where $\rho := (\rho_n)$ is a positive sequence such that

$$\sum_{n=0}^{\infty}\rho_n\gamma(r,n) < +\infty.$$

(ii) $G(\cdot,0,0) \in l_1(\mathbb{Z}_+;X)$.

Remark 6.3.1. If $\rho := (\rho_n)$ is a positive sequence such that

$$\lim_{n\to\infty}\frac{\rho_{n+1}}{\rho_n} = \beta_0 < \min\{1,1/r\},$$

then $\sum_{n=0}^{\infty}\rho_n\gamma(r,n) < +\infty$. For a concrete example we can consider $G(n,z) = \beta_0^n Q(z)$, where

$$||Q(z) - Q(w)||_{X \times X} \leq C ||z - w||_{X \times X}.$$

The next result ensures the existence of solutions whose second discrete r-derivative is in l_p.

Theorem 6.3.2. *Assume that (5.5.11) is fulfilled. Let X be a UMD space and $T \in \mathcal{B}(X)$ be an analytic operator and assume that the following conditions hold:*

(a) *Assumption (A_4) is fulfilled.*
(b) *The set*

$$\mathcal{M} = \left\{ M(z) := (z - r)^2 \left((z - r)^2 - r^2(I - T) \right)^{-1} : |z| = r\alpha, \ z \neq \alpha r \right\}$$

is R-bounded.

Then, there is a unique solution $x = (x_n)$ of (6.3.1) such that $(\Delta_r^2 x_n)$, $((I - T)x_n)$ belong to $l_p(\mathbb{Z}_+; X)$. Moreover, one has the following a priori estimates for the solution:

$$\sup_{n \in \mathbb{Z}_+} \left(\frac{1}{\gamma(r,n)} \left(||x_n||_X + ||\Delta_r x_n||_X \right) \right) \leq ||G(\cdot, 0, 0)||_1 ||\rho. \gamma(r, \cdot)||_1^{-1} e^{2A(r)},$$

$$(6.3.3)$$

and

$$||\Delta_r^2 x||_p + ||r^2(I - T)x||_p \leq C(r) ||G(\cdot, 0, 0)||_1 e^{2A(r)}, \quad 1 < p < +\infty, \ (6.3.4)$$

where $C(r)$ is a constant depending on r and $A(r)$ is given in (6.3.2).

The proof of Theorem 6.3.2 is based on ideas contained in the proof of Theorem 6.2.2 (see Castro and Cuevas [37] for details).

Example 6.3.3. Assume that (5.5.11) is fulfilled. Let H be a Hilbert space and $T \in \mathcal{B}(H)$ be an analytic operator such that the set $\mathcal{M}$ in (b) of the previous theorem is bounded. Consider the following semilinear problem in H:

$$\begin{cases} \Delta_r^2 x_n - r^2(I - T)x_n = v^n f(ax_n + b\Delta_r x_n), \ n \in \mathbb{Z}_+, \\ \\ x_0 = x_1 = 0, \end{cases} \quad (6.3.5)$$

where f is defined and satisfies a Lipschitz condition with constant δ, $0 < v < \min\{1, 1/r\}$ and $a, b \in \mathbb{R}$. Then, by Theorem 6.3.2, there is a unique solution $x = (x_n)$ of (6.3.5) such that $(\Delta_r^2 x_n)$, $((I - T)x_n)$ belong to $l_p(\mathbb{Z}_+; H)$.

In particular, take $T = I$ the identity operator. If f is defined and satisfies a Lipschitz condition in $\mathbb{R}$, we obtain the following scalar result.

Corollary 6.3.4. *Assume that (5.5.11) is fulfilled and let $q = (q_n)$ be a complex sequence such that $\sum_{n=0}^{\infty} |q_n| \gamma(r,n) < +\infty$. Then the problem*

$$\begin{cases} \Delta_r^2 x_n = q_n f(x_n), \ n \in \mathbb{Z}_+, \\ \\ x_0 = x_1 = 0, \end{cases}$$

has a unique solution $x = (x_n)$ such that $(\Delta_r^2 x_n)$ belongs to $l_p(\mathbb{Z}_+; \mathbb{R})$.

Example 6.3.5. We consider the following scalar evolution problem (see [75, Example 7.41]):

$$\begin{cases} y_{n+2} - \dfrac{3}{2} y_{n+1} + \dfrac{1}{2} y_n = \dfrac{\exp^{-n}}{(1 + y_n^2)}, \ n \in \mathbb{Z}_+, \\ \\ y_0 = y_1 = 0. \end{cases} \qquad (6.3.6)$$

By Example 6.3.3, there is a unique solution $y = (y_n)$ of (6.3.6) such that $y = (y_n)$ belongs to $l_p(\mathbb{Z}_+; \mathbb{R})$. In fact, take $T = \frac{8}{9}I$, $v = \frac{1}{e}$, $r = \frac{3}{4}$, $a = 1$, $b = 0$, $f(x) = \frac{1}{(1+x^2)}$, $\delta = 1$ in Example 6.3.3.

The next corollary was inspired by Kunstmann and Weis' $T-$small perturbation results (see [126, Theorem 1]).

Corollary 6.3.6. *Under the conditions of Theorem 6.3.2, if $S \in \mathcal{B}(X)$ is a $T-$small operator satisfying $\|Sx\|_X \le a\|Tx\|_X$ for all $x \in X$ with $0 < a < 2/(\|T\|R(\mathcal{M}))$, where $\mathcal{M}$ is the set in (b) of Theorem 6.3.2 and $R(\mathcal{M})$ denotes the $R-$bound of $\mathcal{M}$ (see Definition 2.2.1), then there is a unique solution $x = (x_n)$ of the equation*

$$\begin{cases} \Delta_r^2 x_n - r^2(I - T)x_n + r^2 S x_n = G(n, x_n, \Delta x_n), \ n \in \mathbb{Z}_+, \\ \\ x_0 = x_1 = 0, \end{cases} \qquad (6.3.7)$$

such that $(\Delta_r^2 x_n)$ and $(x_n - (T + S)x_n)$ belong to $l_p(\mathbb{Z}_+; X)$.

Proof. Since $\mathcal{M}$ is a $R-$bounded subset of $\mathcal{B}(X)$, $\mathcal{M}$ is bounded and

$$\sup_{\Phi \in \mathcal{M}} \|\Phi\| \le R(\mathcal{M}).$$

Thus, for $x \in X$ we obtain

$$\|SR((z-r)^2, R^2(I-T))x\|_X$$

$$\le \frac{a}{2r^2} \|T\| \|M(z)\| \|x\|_X$$

$$\le \frac{a}{2r^2} \|T\| R(\mathcal{M}) \|x\|_X.$$

Hence $I + r^2 SR((z - r)^2, r^2(I - T))$ is invertible and

$$
\begin{aligned}
(z - r)^2 &R((z - r)^2, r^2(I - (T + S))) \\
&= M(z)\left(I + \frac{r^2}{(z - r)^2} SM(z)\right)^{-1} \\
&= M(z) \sum_{k=0}^{\infty} \frac{r^{2k}}{(z - r)^{2k}} (SM(z))^k.
\end{aligned}
$$

Now, one can apply a similar argument as in [22, Lemma 3.4] to infer that

$$
\left\{(z - r)^2 R((z - r)^2, r^2(I - (T + S))) : |z| = \alpha r, z \neq \alpha r\right\},
$$

is R-Bounded. Finally, one uses Theorems 5.6.3 and 6.3.2 to conclude the proof of the corollary.   □

6.4   Local Perturbations

In the process of obtaining our next result, we will require the following assumption.

Assumption $(A_2)^*$:   The following conditions hold:

(i) The function $f(n, z)$ is locally Lipschitz with respect to $z \in X \times X$, i.e., for each positive number R, for all $n \in \mathbb{Z}_+$ and $z, w \in X \times X$, $\|z\|_{X \times X} \leq R, \|w\|_{X \times X} \leq R$, we have

$$
\|f(n, z) - f(n, w)\|_X \leq l(n, R)\|z - w\|_{X \times X},
$$

where $l : \mathbb{Z}_+ \times [0, \infty) \longrightarrow [0, \infty)$ is a nondecreasing function with respect to the second variable.

(ii) There is a positive number a such that

$$
\sum_{n=0}^{\infty} \ell(n, a) < +\infty.
$$

(iii) $f(\cdot, 0, 0) \in \ell_1(\mathbb{Z}_+; X)$.

We have the following local version of Theorem 6.1.1.

Theorem 6.4.1. *Suppose that the following conditions are satisfied:*

(a)* *Assumption $(A_2)^*$ holds.*
(b)* *Equation (5.1.1) has discrete maximal l_p-regularity.*

Then, there is a positive constant $m \in \mathbb{N}$ and a unique bounded solution x_n of (6.1.1) for $n > m$ such that $x_n = 0$ if $0 \leq n \leq m$ and the sequence $(\Delta^2 x_n)$ belongs to $\ell_p(\mathbb{Z}_+; X)$. Moreover, we get

$$\|x\|_\infty + \|\Delta^2 x\|_p \leq a, \tag{6.4.1}$$

where a is the constant of condition $(A_2)^*$-(ii).

Corollary 6.4.2. *Let $B_i : X \times X \longrightarrow X$, $i = 1,2$ are two bounded bilinear operators; $y \in \ell_1(\mathbb{Z}_+; X)$ and $\alpha, \beta \in \ell_1(\mathbb{Z}_+; \mathbb{R})$. In addition suppose that (5.1.1) has discrete maximal l_p-regularity. Then, there is a unique bounded solution x such that $(\Delta^2 x) \in l_p(\mathbb{Z}_+; X)$ for the equation*

$$x_{n+2} - 2x_{n+1} + T x_n = y_n + \alpha_n B_1(x_n, x_n) + \beta_n B_2(\Delta x_n, \Delta x_n).$$

6.5   Local Perturbations for the Harmonic Oscillator

In the next result we suppose that the following condition is satisfied.

Assumption $(A_3)^*$:   (i)*   The function $f(n, z)$ is locally Lipschitz with respect to $z \in X \times X$ satisfying $(A_2)^*$ (i) and there is a positive number a such that

$$\sum_{n=0}^{\infty} \ell(n, 2^{2n} a) 2^{2n} < +\infty.$$

(ii)* $f(\cdot, 0, 0) \in \ell_1(\mathbb{Z}_+; X)$.

We have the following local version of Theorem 6.2.2.

Theorem 6.5.1. *Suppose that (5.3.1) has discrete maximal l_p-regularity. Then, there is a positive constant $m \in \mathbb{N}$ and a unique solution $x = (x_n)$ of (6.2.1) for $n > m$ such that $x_n = 0$ if $0 \leq n \leq m$ and the sequence $(\Delta^2 x_n)$ belongs to $\ell_p(\mathbb{Z}_+; X)$ with $\|\Delta^2 x\|_p \leq a$ where a is the constant of condition $(A_3)^*$-(i)*.*

Proof. Let β in $(0, 1)$. There are n_1 and n_2 in $\mathbb{N}$ such that

$$(1 + \|\mathcal{K}_B\|) \sum_{j=n_1}^{\infty} \|f(j, 0, 0)\|_X \leq \beta a, \tag{6.5.1}$$

and

$$\mathcal{T} := \beta + (1 + \|\mathcal{K}_B\|) \sum_{j=n_2}^{\infty} \ell(j, 2^{2j} a) 2^{2j} < 1. \tag{6.5.2}$$

Let V be a sequence in $\mathcal{H}_m^{2,p}[a]$, with $m = \max\{n_1, n_2\}$ [5]. An argument involving condition $(A_3)^*$ and (6.2.6) [6] shows that the sequence

$$g_n := \begin{cases} 0, & \text{if } 0 \le n \le m, \\ f(n, V_n, \Delta V_n), & \text{if } n > m, \end{cases} \tag{6.5.3}$$

belongs to ℓ_p. In fact, we have

$$\|g\|_p = \Big(\sum_{n=m+1}^{\infty} \|f(n, V_n, \Delta V_n)\|_X^p \Big)^{1/p}$$

$$\le \sum_{n=m+1}^{\infty} l(n, 2^{2n}a)\|(V_n, \Delta V_n)\|_{X \times X} + \sum_{n=m+1}^{\infty} \|f(n, 0, 0)\|_X$$

$$\le \sum_{n=m+1}^{\infty} l(n, 2^{2n}a)2^{2n}a + \sum_{n=m+1}^{\infty} \|f(n, 0, 0)\|_X.$$

By the discrete maximal regularity, the Cauchy problem (6.2.8) with (g_n) defined as in (6.5.3) has a unique solution (z_n) such that $(\Delta^2 z_n) \in l_p(\mathbb{Z}_+; X)$, which is given by

$$z_n = [\Upsilon V]_n = \begin{cases} 0, & \text{if } 0 \le n \le m, \\ \displaystyle\sum_{k=0}^{n-1-m} B(k) f(n-1-k, V_{n-1-k}, \Delta V_{n-1-k}), & \text{if } n \ge m+1. \end{cases}$$

$$\tag{6.5.4}$$

We will prove that ΥV belongs to $\mathcal{H}_m^{2,p}[a]$. In fact,[7] since $B(0) = 0$ and $B(1) = I$ (see (5.7.1)) we get for $n \ge m$ the following identities:

$$\Delta^2[\Upsilon V]_n = B(1) f(m, V_m, \Delta V_m) = f(m, V_m, \Delta V_m),$$

---

[5] We denote by $\mathcal{H}_m^{2,p}$ the Banach space of all sequences $V = (V_n)$ such that $V_n = 0$ if $0 \le n \le m$, and $\Delta^2 V \in \ell_p(\mathbb{Z}_+; X)$ equipped with the norm $\|\|V\|\| = \|\Delta^2 V\|_p$. For $\lambda > 0$, denote by $\mathcal{H}_m^{2,p}[\lambda]$ the ball $\|\|V\|\| \le \lambda$ in $\mathcal{H}_m^{2,p}$.

[6] We observe that $\|(V_n, \Delta V_n)\|_{X \times X} \le 2^{2n}\|\|V\|\| \le 2^{2n}a$.

[7] We note that $B(0) = 0$, $B(1) = I$; hence

$$\Delta[\Upsilon V]_n = \sum_{k=0}^{n-1-m} (B(k+1) - B(k)) f(n-1-k, V_{n-1-k}, \Delta V_{n-1-k}), \ n \ge m+1.$$

and

$$\Delta^2[\Upsilon V]_n = \Delta[\Upsilon V]_{n+1} - \Delta[\Upsilon V]_n$$
$$= \sum_{j=0}^{n-m} (B(j+1) - B(j)) f(n-j, V_{n-j}, \Delta V_{n-j})$$

$$- \sum_{j=0}^{n-1-m} (B(j+1) - B(j)) f(n-1-j, V_{n-1-j}, \Delta V_{n-1-j})$$

$$= (B(1) - B(0)) f(n, V_n, \Delta V_n)$$
$$+ \sum_{j=0}^{n-1-m} (B(j+2) - B(j+1)) f(n-1-j, V_{n-1-j}, \Delta V_{n-1-j})$$

$$- \sum_{j=0}^{n-1-m} (B(j+1) - B(j)) f(n-1-j, V_{n-1-j}, \Delta V_{n-1-j})$$

$$= f(n, V_n, \Delta V_n) + \sum_{j=0}^{n-1-m} (B(j+2) - 2B(j+1)$$

$$+ B(j)) f(n-1-j, V_{n-1-j}, \Delta V_{n-1-j})$$

$$= f(n, V_n, \Delta V_n) + \sum_{j=0}^{n-1-m} \Delta^2 B(j) f(n-1-j, V_{n-1-j}, \Delta V_{n-1-j})$$

$$= f(n, V_n, \Delta V_n)$$

$$+ \sum_{j=0}^{n-1-m} (I - T) B(j+1) f(n-1-j, V_{n-1-j}, \Delta V_{n-1-j})$$

$$= f(n, V_n, \Delta V_n) + \sum_{k=1}^{n-m} (I - T) B(k) f(n-k, V_{n-k}, \Delta V_{n-k}).$$

Hence

$$\|\Delta^2 \Upsilon V\|_p = \left(\|f(m, V_m, \Delta V_m)\|_X^p + \sum_{n=m+1}^{\infty} \|\Delta^2[\Upsilon V]_n\|_X^p \right)^{1/p}$$

$$\leq \|f(m, V_m, \Delta V_m)\|_X + \Big(\sum_{n=m+1}^{\infty} \Big\| f(n, V_n, \Delta V_n)$$

$$+ \sum_{k=1}^{n-m} (I - T)B(k) f(n - k, V_{n-k}, \Delta V_{n-k}) \Big\|_X^p \Big)^{1/p}$$

$$\leq \|f(m, V_m, \Delta V_m)\|_X + \Big(\sum_{n=m+1}^{\infty} \|f(n, V_n, \Delta V_n)\|_X^p \Big)^{1/p}$$

$$+ \Big(\sum_{j=1}^{\infty} \Big\| \sum_{k=1}^{j} (I - T)B(k) f(j - k + m, V_{j-k+m}, \Delta V_{j-k+m}) \Big\|_X^p \Big)^{1/p}$$

$$\leq \sum_{n=m}^{\infty} \|f(n, V_n, \Delta V_n)\|_X$$

$$+ \|\mathcal{K}_B\| \Big(\sum_{j=1}^{\infty} \|f(j + m, V_{j+m}, \Delta V_{j+m})\|_X^p \Big)^{1/p}$$

$$\leq (1 + \|\mathcal{K}_B\|) \sum_{n=m}^{\infty} \|f(n, V_n, \Delta V_n)\|_X.$$

Therefore

$$\|\|\Upsilon V\|\| \leq (1 + \|\mathcal{K}_B\|) \sum_{j=m}^{\infty} \ell(j, 2^{2j}a) 2^{2j} a + (1 + \|\mathcal{K}_B\|) \sum_{j=m}^{\infty} \|f(j, 0, 0)\|_X$$

$$\leq (\mathcal{T} - \beta)a + \beta a \leq a,$$

proving that ΥV belongs to $\mathcal{H}_m^{2,p}[a]$.

For all V and W in $\mathcal{H}_m^{2,p}[a]$, we can prove that

$$\|\|\Upsilon V - \Upsilon W\|\| \leq (1 + \|\mathcal{K}_B\|) \sum_{j=m}^{\infty} \ell(j, 2^{2j}a) 2^{2j} \|\|V - W\|\|$$

$$= (\mathcal{T} - \beta) \|\|V - W\|\|.$$

Since $(\mathcal{T} - \beta) < 1$, Υ is a $(\mathcal{T} - \beta)$-contraction. This completes the proof of the theorem. $\square$

6.6   Local Perturbations on Weighted Spaces

We require the following assumption.

Assumption $(A_4)^*$:   The following conditions hold:

(a)* The function $G(n, z)$ is locally Lipschitz with respect to $z \in X \times X$, that is, for each positive number R, for all $n \in \mathbb{Z}_+$ and $z, w \in X \times X$, $||z||_{X \times X} \leq R$, $||w||_{X \times X} \leq R$, we have

$$||G(n, z) - G(n, w)||_X \leq l(n, R)||z - w||_{X \times X},$$

where $l : \mathbb{Z}_+ \times [0, \infty) \longrightarrow [0, \infty)$ is a nondecreasing function with respect to the second variable.

(b)* There is a positive number a such that

$$\sum_{n=0}^{\infty} \ell(n, \Theta(r)\gamma(r, n)a)\gamma(r, n) < +\infty,$$

where $\Theta(r)$ and $\gamma(r, n)$ are given by (5.5.15) and (5.5.16), respectively.

(c)* $G(\cdot, 0, 0) \in \ell_1(\mathbb{Z}_+; X)$.

We have the following local version of Theorem 6.3.2.

Theorem 6.6.1. *Assume that (5.5.11) is fulfilled. Let X be a UMD space and $T \in \mathcal{B}(X)$ be an analytic operator and assume that conditions $(A_4)^*$ and (b) of Theorem 6.3.2 hold. Then, there is a positive constant $m \in \mathbb{Z}_+$ and a unique solution $x = (x_n) \in MR_p(\mathbb{Z}_+; X)$ of (6.3.1) for $n > m$ such that $x_1 = \cdots = x_m = 0$. Moreover, one has*

$$||\Delta_r^2 x||_p + ||r^2(I - T)x||_p \leq a, \tag{6.6.1}$$

where a is a constant of condition $(A_4)^*$-(b).*

Proof. Let $\beta \in (0, 1)$. Using (c)* and (b)* respectively, there is $m \in \mathbb{Z}_+$ such that

$$2(1 + ||\mathcal{K}^r||_{\mathcal{B}(l_p(\mathbb{Z}_+; X))}) \sum_{n=m}^{\infty} ||G(n, 0, 0)||_X \leq \beta a,$$

and

$$\mathcal{T} := \beta + 2(1 + ||\mathcal{K}^r||_{\mathcal{B}(l_p(\mathbb{Z}_+; X))})\Theta(r) \sum_{n=m}^{\infty} \ell(n, \Theta(r)\gamma(r, n)a)\gamma(r, n) < 1,$$

where $\mathcal{K}^r$ is the operator given in Definition 5.5.3.

We denote by $\mathcal{W}_m$ the Banach space of all sequences $y = (y_n)$ such that $y_1 = \cdots = y_m = 0$, and $\Delta_r^2 y$, $(I - T)y$ belong to $l_p(\mathbb{Z}_+; X)$ equipped with the norm $\|\cdot\|_{MR_p}$ given by (5.6.2). We denote by $\mathcal{W}_m[a]$ the ball $\|y\|_{MR_p} \leq a$ in $\mathcal{W}_m$. Let y be a sequence in $\mathcal{W}_m[a]$. We can see that the sequence

$$f_n := \begin{cases} 0, & \text{if } 0 \leq n \leq m, \\ G(n, y_n, \Delta_r y_n), & \text{if } n > m, \end{cases}$$

belongs to $l_p(\mathbb{Z}_+; X)$. By Theorem 5.6.3, the Cauchy problem

$$\begin{cases} \Delta_r^2 z_n - r^2(I - T)z_n = f_n, \ n \in \mathbb{Z}_+, \\ z_0 = z_1 = 0, \end{cases}$$

has a unique solution (z_n) such that $(\Delta_r^2 z_n)$ and $((I - T)z_n)$ belong to $l_p(\mathbb{Z}_+; X)$, which is given by

$$z_n = [\mathcal{N}y]_n = \begin{cases} 0, & 0 \leq n \leq m, \\ \displaystyle\sum_{k=0}^{n-1-m} r^{k-1}S(k)G(n-1-k, y_{n-1-k}, \Delta_r y_{n-1-k}), & n \geq m+1. \end{cases}$$

Let y be a sequence in $\mathcal{W}_m[a]$. We see that $\mathcal{N}y$ belongs to $\mathcal{W}_m[a]$. In fact, we have

$$\|\mathcal{N}y\|_{MR_p} \leq 2(1 + \|\mathcal{K}^r\|_{\mathcal{B}(l_p(\mathbb{Z}_+;X))})\Theta(r) \sum_{j=m}^{\infty} \ell(j, \Theta(r)\gamma(r, j)a)\gamma(r, j)a$$

$$+ 2(1 + \|\mathcal{K}^r\|_{\mathcal{B}(l_p(\mathbb{Z}_+;X))}) \sum_{j=m}^{\infty} \|G(j, 0, 0)\|_X$$

$$\leq (1 - \beta)a + \beta a = a.$$

On the other hand, for all y^1 and y^2 in $\mathcal{W}_m[a]$, we have

$$\|\mathcal{N}y^1 - \mathcal{N}y^2\|_{MR_p} \leq 2(1 + \|\mathcal{K}^r\|_{\mathcal{B}(l_p(\mathbb{Z}_+;X))}) \sum_{n=m}^{\infty} \ell(n, \Theta(r)\gamma(r, n)a)$$

$$\times \|((y^1 - y^2)_n, \Delta_r(y^1 - y^2)_n)\|_{X \times X}$$

$$\leq (\mathcal{T} - \beta)\|y^1 - y^2\|_{MR_p}.$$

Hence $\mathcal{N}$ is a $(\mathcal{T} - \beta)$-contraction. This completes the proof of the theorem. $\square$

Corollary 6.6.2. *Assume that (5.5.11) and* $(A_4)^*$ *are fulfilled. Let H be a Hilbert space and $T \in B(H)$ be an analytic operator. If*

$$\sup_{|z|=\alpha r,\, z\neq\alpha r} \left\| (z-r)^2 R((z-r)^2, r^2(I-T)) \right\| < +\infty,$$

then there is a positive constant $m \in \mathbb{Z}_+$ and a unique solution $x = (x_n) \in MR_p(\mathbb{Z}_+; H)$ of (6.3.1) for $n \geq m$ such that $x_1 = \cdots = x_m = 0$.

Example 6.6.3. Assume that (5.5.11) is fulfilled. Let p be a number such that $1 < p < \infty$ and $y \in l_1(\mathbb{Z}_+; \mathbb{R})$. Then the following discrete evolution equation:

$$\begin{cases} \Delta_r^2 x_n = y_n + \dfrac{1}{n^p} |\Delta_r x_n|^2, \; n \in \mathbb{Z}_+, \\[2mm] x_0 = x_1 = 0, \end{cases} \tag{6.6.2}$$

has a unique bounded solution $x = (x_n)$ such that $(\Delta_r^2 x_n)$ belongs to $l_p(\mathbb{Z}_+; \mathbb{R})$.

6.7 Comments

Theorem 6.1.1 is established by Cuevas and Lizama in [55]. Theorem 6.5.1 is from [52].

In the continuous case, it is well known that the study of maximal regularity is very useful for treating semilinear and quasilinear problems (see,e.g., Amann [6], Denk–Hieber and Prüss [65], Clément–Londen–Simonett [46], the survey by Arendt [10], and the bibliography therein).

Chapter 7
Applications

In this chapter we present several different types of applications concerning semilinear difference equations. The important problem of finding the practical criterion for R-boundedness is studied in Sect. 7.3. In addition, boundedness and asymptotic behavior of solutions are analyzed. Finally, a criterion for the boundedness of semilinear functional difference equations with infinite delay is presented.

7.1 Semilinear Difference Equations

Let $f = (f_n) \in l_p(\mathbb{Z}_+; X)$. We begin this section with an application to the existence of solutions to the following semilinear evolution problem:

$$\Delta_r^2 x - r^2(I - T)x = G(x) + \rho f, \tag{7.1.1}$$

where G is a Frechét differentiable function and $\rho > 0$ is a small parameter. We note that the above problem corresponds to a discrete version of an integrodifferential equation initially considered in [118].

The following theorem is a consequence of Theorem 5.6.3 and the implicit function theorem.

Theorem 7.1.1. *Let X be a UMD space and let $T \in B(X)$ be an analytic operator. Assume that (5.5.11) is fulfilled and assume that*

(i) *the set*

$$\left\{ (z - r)^2 \left((z - r)^2 - r^2(I - T) \right)^{-1} : |z| = \alpha r, z \neq \alpha r \right\}$$

is R-bounded and

R.P. Agarwal et al., *Regularity of Difference Equations on Banach Spaces*, DOI 10.1007/978-3-319-06447-5_7, © Springer International Publishing Switzerland 2014

(ii) G maps $MR_p(\mathbb{Z}_+; X)$ into $l_p(\mathbb{Z}_+; X)$; $G(0) = 0$; G is a continuously (Frechét) differentiable function at $x = 0$ and $G'(0) = 0$.

Then, there exists $\rho^* > 0$ such that the (7.1.1) is solvable for each $\rho \in [0, \rho^*)$, with solution $x_\rho := (x_n) \in MR_p(\mathbb{Z}_+; X)$, [see (5.6.1)].

Proof. Define the operator $L : l^2_{p,r}(\mathbb{Z}_+; X) \cap l_{p,I-T}(\mathbb{Z}_+; X) \to l_p(\mathbb{Z}_+; X)$ by

$$L(x) = \Delta_r^2 x - r^2(I - T)x.$$

We consider for $\rho \in (0, 1)$ the one parameter family

$$H[x, \rho] := -L(x) + G(x) + \rho f.$$

Keeping in mind that $G(0) = 0$ we see that $H[0, 0] = 0$. Also, by hypothesis, H is continuously differentiable at $(0, 0)$. We observe that L is an isomorphism. In fact, by uniqueness L is injective. By Theorem 5.6.3, L is surjective. By definition of the norm in (5.6.2), L is bounded. Now, the claim follows by the Open Mapping Theorem. Hence the partial Frechét derivative $H^1_{(0,0)} = -L$ is invertible. The conclusion of the theorem now follows from the implicit function theorem (see [87, Theorem 17.6]). $\square$

In the case $T = I$ and $X = \mathbb{R}$ we obtain the following scalar result.

Corollary 7.1.2. *Assume that (5.5.11) is fulfilled; let $f = (f_n)$ be in $l_p(\mathbb{Z}_+; \mathbb{R})$. Assume that G is as in the preceding theorem. Then, there is $\rho^* > 0$ such that the discrete-time evolution equation*

$$\Delta_r^2 x = G(x) + \rho f, \tag{7.1.2}$$

is solvable for each $\rho \in [0, \rho^*)$, with solution $x_\rho := (x_n) \in MR_p(\mathbb{Z}_+; \mathbb{R})$.

Next, we exhibit a concrete example of mapping G which satisfies the conditions of Theorem 7.1.1.

Example 7.1.3. Let $T = I$ and $X = \mathbb{R}$ and let G be a map from $MR_p(\mathbb{Z}_+; \mathbb{R})$ into $l_p(\mathbb{Z}_+; \mathbb{R})$ defined by

$$G(h)(n) = (\Delta_r^2 h_n)^2.$$

For $h \in MR_p(\mathbb{Z}_+; \mathbb{R})$, we have

$$\|G(h)\|_{l_p(\mathbb{Z}_+; \mathbb{R})} = \left(\sum_{n=0}^{\infty} |\Delta_r^2 h_n|^p |\Delta_r^2 h_n|^p \right)^{\frac{1}{p}}$$

$$\leq \|\Delta_r^2 h\|_{l_\infty(\mathbb{Z}_+; \mathbb{R})} \left(\sum_{n=0}^{\infty} |\Delta_r^2 h_n|^p \right)^{\frac{1}{p}}$$

$$= \|\Delta_r^2 h\|_{l_\infty(\mathbb{Z}_+; \mathbb{R})} \cdot \|\Delta_r^2 h\|_{l_p(\mathbb{Z}_+; \mathbb{R})},$$

whence G is well defined.[1]

On the other hand we get

$$G'(x)(h)(n) = \lim_{t\to 0} \frac{1}{t}\Big((\Delta_r^2 x_n + t\Delta_r^2 h_n)^2 - (\Delta_r^2 x_n)^2\Big)$$

$$= \lim_{t\to 0} \frac{1}{t}\Big(2t\Delta_r^2 x_n \cdot \Delta_r^2 h_n + t^2(\Delta_r^2 h_n)^2\Big)$$

$$= 2\Delta_r^2 x_n \cdot \Delta_r^2 h_n.$$

Therefore[2]

$$G'(x)(h) = 2\Delta_r^2 x \cdot \Delta_r^2 h, \quad x, h \in MR_p(\mathbb{Z}_+; \mathbb{R}),$$

so $G'(0) = 0$.

We can infer that[3]

$$\|G'(x) - G'(x^0)\| \le 2\|x - x^0\|_{MR_p}, \quad x, x^0 \in MR_p(\mathbb{Z}_+; \mathbb{R}),$$

then G is a continuously (Frechét) differentiable function.

Example 7.1.4. Assume that (5.5.11) is fulfilled. Let μ be a positive integer and $f = (f_n) \in l_p(\mathbb{Z}_+; \mathbb{R})$. Then, there is $\rho^* > 0$ such that the discrete-time evolution problem

$$\begin{cases} \Delta_r^2 x_n = [\Delta_r^2 x_n]^\mu + \rho f_n, \ n \in \mathbb{Z}_+, \\[2mm] x_0 = x_1 = 0, \end{cases} \tag{7.1.3}$$

is solvable for each $\rho \in [0, \rho^*)$, with solution $x_\rho := (x_n) \in MR_p(\mathbb{Z}_+; \mathbb{R})$.

---

[1] We observe that $G(0) = 0$.

[2] Put $r(h) = G(x+h) - G(x) - 2\Delta_r^2 x \cdot \Delta_r^2 h$, then $\frac{\|r(h)\|_p}{\|h\|_{MR_p}} \le \|h\|_{MR_p}$. Hence $\lim_{h\to 0} \frac{r(h)}{\|h\|_{MR_p}} = 0$.

[3] We observe that

$$\|G'(x)h - G'(x^0)h\|_p = 2\Big(\sum_{n=0}^{\infty} |\Delta_r^2 x_n - \Delta_r^2 x_n^0|^p |\Delta_r^2 h_n|^p\Big)^{\frac{1}{p}}$$

$$\le 2\|\Delta_r^2(x - y)\|_p \|\Delta_r^2 h\|_\infty$$

$$\le 2\|x - x^0\|_{MR_p} \|h\|_{MR_p}.$$

Proof. Let G be the map associated to (7.1.3). We observe that $G(0) = 0$. For $h \in MR_p(\mathbb{Z}_+; \mathbb{R})$, we have

$$||G(h)||_{l_p(\mathbb{Z}_+;\mathbb{R})} = \left(\sum_{n=0}^{\infty} |\Delta_r^2 h_n|^{(\mu-1)p} |\Delta_r^2 h_n|^p \right)^{\frac{1}{p}}$$

$$\leq ||\Delta_r^2 h||_{l_\infty(\mathbb{Z}_+;\mathbb{R})}^{\mu-1} \cdot ||\Delta_r^2 h||_{l_p(\mathbb{Z}_+;\mathbb{R})},$$

hence $G(h) \in l_p(\mathbb{Z}_+; \mathbb{R})$.

On the other hand, we note that

$$\frac{1}{t}(G(x+th)(n) - G(x)(n))$$

$$= \frac{1}{t}((\Delta_r^2 x_n + t\Delta_r^2 h_n)^\mu - (\Delta_r^2 x_n)^\mu)$$

$$= (\Delta_r^2 h_n) \sum_{j=0}^{\mu-1} (\Delta_r^2 x_n + t\Delta_r^2 h_n)^{\mu-1-j} (\Delta_r^2 x_n)^j.$$

Therefore

$$\lim_{t \to 0} \frac{1}{t}(G(x+th)(n) - G(x)(n)) = \mu(\Delta_r^2 x_n)^{\mu-1}(\Delta_r^2 h_n).$$

Now putting $r(h) = G(x+h)(n) - G(x)(n) - \mu(\Delta_r^2 x)^{\mu-1}(\Delta_r^2 h)$. From the following identity:

$$r(h) = (\Delta_r^2 h)\left(\sum_{j=0}^{\mu-1} (\Delta_r^2 x + \Delta_r^2 h)^{\mu-1-j}(\Delta_r^2 x)^j - \mu(\Delta_r^2 x)^{\mu-1} \right)$$

$$= (\Delta_r^2 h) \sum_{j=0}^{\mu-1} \left((\Delta_r^2 x + \Delta_r^2 h)^{\mu-1-j}(\Delta_r^2 x)^j - (\Delta_r^2 x)^{\mu-1} \right)$$

$$= (\Delta_r^2 h) \sum_{j=0}^{\mu-1} \left((\Delta_r^2 x + \Delta_r^2 h)^{\mu-1-j} - (\Delta_r^2 x)^{\mu-1-j} \right)(\Delta_r^2 x)^j.$$

We get

$$\lim_{h \to 0} \frac{r(h)}{||h||_{MR_p}} = 0.$$

Consequently

$$G'(x)(h) = \mu(\Delta_r^2 x)^{\mu-1} \cdot (\Delta_r^2 h), \quad x, h \in MR_p(\mathbb{Z}_+; \mathbb{R}).$$

We observe that G is continuously (Frechét) differentiable. In fact, we can deduce the following estimate:

$$||G'(x)h - G'(x^0)h||_p \leq \mu||x - x^0||_{MR_p}||h||_{MR_p}, \quad x, x^0, h \in MR_p(\mathbb{Z}_+; \mathbb{R}),$$

that is,

$$||G'(x) - G'(x^0)|| \leq \mu||x - x^0||_{MR_p}, \quad x, x^0 \in MR_p(\mathbb{Z}_+; \mathbb{R}).$$

By Corollary 7.1.2 there is $\rho^* > 0$ such that the discrete-time evolution problem (7.1.3) is solvable for each $\rho \in [0, \rho^*)$, with solution $x_\rho := (x_n) \in MR_p(\mathbb{Z}_+; \mathbb{R})$. This finishes the discussion of Example 7.1.4. $\qquad\square$

7.2 An Arendt–Duelli-Type Theorem

The following application (see Theorem 7.2.1) corresponds to an Arendt–Duelli-type theorem for evolution equations (see [11, Theorem 4.2]).

Let $T(0)$ be a bounded operator in X and assume here and subsequently until the end of this subsection that (5.5.11) is fulfilled. We consider the maximal regularity space

$$MR_p := l_{p,r}^2(\mathbb{Z}_+; X) \cap l_{p,I-T(0)}(\mathbb{Z}_+; X)$$

equipped with the norm

$$||x||_{MR_p} := ||\Delta_r^2 x||_p + ||r^2(I - T(0))x||_p.$$

For $\rho > 0$ denote by

$$B[\rho] := \{u \in X : ||r^2(I - T(0))u||_X \leq \rho\}$$

the closed ball of radius ρ. Let $\rho_0 > 0$ and $T : B[\rho_0] \to \mathcal{B}(X)$ be a function such that

$$\|T(u) - T(v)\|_{B(X)} \leq L\|(I - T(0))(u - v)\|_X, \qquad (7.2.1)$$

for all $u, v \in B[\rho_0]$ and some constant L.

We assume that the discrete evolution problem

$$\begin{cases} \Delta_r^2 x_n - r^2(I - T(0))x_n = g_n, \ n \in \mathbb{Z}_+, \\ \\ x_0 = x_1 = 0, \end{cases} \qquad (7.2.2)$$

is well posed; that is, for all $g = (g_n) \in l_p(\mathbb{Z}_+; X)$, there is a unique solution $x = (x_n) \in MR_p$ of (7.2.2). Denote by M the norm of the solution operator

$$g \in l_p(\mathbb{Z}_+; X) \to x \in MR_p.$$

Let $G : \mathbb{Z}_+ \times B[\rho_0] \times B[\rho_0] \to X$ be a function such that

$$\|G(n, z_1, w_1)\|_X \leq h_1(n)r^2 \Big(\|(I - T(0))z_1\|_X + \|(I - T(0))w_1\|_X \Big), \quad (7.2.3)$$

and

$$\|G(n, z_1, w_1) - G(n, z_2, w_2)\|_X$$
$$\qquad\qquad (7.2.4)$$
$$\leq h_2(n)r^2 \Big(\|(I - T(0))(z_1 - z_2)\|_X + \|(I - T(0))(w_1 - w_2)\|_X \Big),$$

for all $n \in \mathbb{Z}_+, z_i, w_i \in B[\rho_0]$, i=1,2, where $h_1, h_2 \in l_p(\mathbb{Z}_+; \mathbb{R})$ such that

$$\|h_1\|_p \leq \big(M(2 + r)\big)^{-1}, \ \|h_2\|_p \leq \big(M(2 + r)\big)^{-1}.$$

We have the following result.[4]

Theorem 7.2.1. *Under above hypotheses there are a radius* $0 < \rho \leq \rho_0$ *and* $\delta > 0$ *such that for each* $f \in l_p(\mathbb{Z}_+; X)$ *with* $\|f\|_p \leq \delta$ *there is a unique* $x = (x_n) \in MR_p$ *with* $\|x\|_{MR_p} \leq \rho$ *satisfying*

$$\begin{cases} \Delta_r^2 x_n - r^2(I - T(x_n))x_n = G(n, x_n, \Delta_r x_n) + f_n, \ n \in \mathbb{Z}_+, \\ \\ x_0 = x_1 = 0. \end{cases} \qquad (7.2.5)$$

---

[4]In the case of initial value problem's existence results for quasilinear equations based on maximal regularity have been obtained by Clément and Li [47]. They proved that a solution exists on some time interval $[0, T]$ for sufficiently small time T. Arendt and Duelli [11] considered solutions on entire line assuming that the inhomogeneity f is sufficiently small. This is also our approach in the Theorem 7.2.1.

Proof. Choose $0 < \rho \leq \rho_0(1+r)^{-1}$ such that

$$LM\Theta(r)\rho + M||h_1||_p(2+r) < 1$$

and

$$2\Theta(r)LM\rho + M||h_2||_p(2+r) < 1,$$

where $\Theta(r)$ is given in $(5.5.15)$. We define

$$\delta := \rho(M^{-1} - \Theta(r)L\rho - (2+r)||h_1||_p).$$

Let f belong to $l_p(\mathbb{Z}_+; X)$ such that $||f||_p \leq \delta$ and let v be in MR_p with $||v||_{MR_p} \leq \rho$. We observe that v_n and $\Delta_r v_n$ belong to $B[\rho_0]$. Indeed, it follows from the following estimates:

$$||r^2(I - T(0))v_n||_X \leq ||r^2(I - T(0))v||_p \leq ||v||_{MR_p} \leq \rho \leq \frac{\rho_0}{1+r} \leq \rho_0$$

and

$$\begin{aligned}
||r^2(I - T(0))\Delta_r v_n||_X &\leq ||r^2(I - T(0))v_{n+1}||_X + r||r^2(I - T(0))v_n||_X \\
&\leq (1+r)||r^2(I - T(0))v||_p \\
&\leq (1+r)||v||_{MR_p} \leq (1+r)\rho \leq \rho_0.
\end{aligned}$$

Hence we can define the function

$$g_n := r^2(T(0) - T(v_n))v_n + G(n, v_n, \Delta_r v_n) + f_n.$$

We claim that $g \in l_p(\mathbb{Z}_+; X)$ and $M||g||_p \leq \rho$. In fact, we have

$$||g_n||_X \leq r^2||(T(0) - T(v_n))||_{\mathcal{B}(X)}||v_n||_X + ||G(n, v_n, \Delta_r v_n)||_X + ||f_n||_X$$

$$\leq L||r^2(I - T(0))v_n||_X||v_n||_X + h_1(n)\big(||r^2(I - T(0))v_n||_X$$

$$+ ||r^2(I - T(0))\Delta_r v_n||_X\big) + ||f_n||_X.$$

From Proposition 5.5.6(i), we have

$$||v_n||_X \leq \Theta(r)||\Delta_r^2 v||_p \leq \Theta(r)||v||_{MR_p} \leq \Theta(r)\rho.$$

Taking into account that $||r^2(I - T(0))v_n||_X \leq \rho$ and $||r^2(I - T(0))\Delta_r v_n||_X$
$\leq (1 + r)\rho$, we infer that

$$M\|g\|_p \leq M\big(L\Theta(r)\rho\|r^2(I - T)v\|_p + (2 + r)\rho\|h_1\|_p + \|f\|_p\big)$$

$$\leq M\big(L\Theta(r)\rho\|v\|_{MR_p} + (2 + r)\rho\|h_1\|_p + \delta\big)$$

$$\leq M\big(\rho M^{-1} - \delta + \delta\big) = \rho.$$

Denote by $\Phi(v) := x$ the solution of (7.2.2) for the inhomogeneity g. Then

$$\|x\|_{MR_p} \leq M\|g\|_p \leq \rho.$$

Thus Φ maps the set

$$\mathcal{A}_\rho := \{v \in MR_p : \|v\|_{MR_p} \leq \rho\}$$

into itself. It remains only to verify that Φ is a strict contraction. In fact, let $x^1 = \Phi(v^1)$, $x^2 = \Phi(v^2)$, $v^1, v^2 \in \mathcal{A}_\rho$. Then $x^1 - x^2$ is solution of (7.2.2) for the inhomogeneity

$$g_n := r^2(T(0) - T(v_n^2))(v_n^1 - v_n^2) + r^2(T(v_n^2) - T(v_n^1))v_n^1 + G(n, v_n^1, \Delta_r v_n^1)$$
$$-G(n, v_n^2, \Delta_r v_n^2).$$

In the same manner as before we estimate

$$\|x^1 - x^2\|_{MR_p} \leq M\|g\|_p \leq M\big(L\Theta(r)\|v^2\|_{MR_p}\|v^1 - v^2\|_{MR_p}$$

$$+ L\Theta(r)\|v^1\|_{MR_p}\|v^1 - v^2\|_{MR_p}$$

$$+ \|h_2\|_p(2 + r)\|v^1 - v^2\|_{MR_p}\big)$$

$$\leq M\big(2\Theta(r)L\rho + \|h_2\|_p(2 + r)\big)\|v^1 - v^2\|_{MR_p}.$$

This clearly implies that Φ is a strict contraction. By Banach's fixed-point theorem there exists a unique fixed point $x \in \mathcal{A}_\rho$ of Φ. □

From Theorems 5.6.3 and 7.2.1 we obtain the following result.

Corollary 7.2.2. *Let X be a UMD space, let $T(0) \in \mathcal{B}(X)$ be an analytic operator, and assume that (5.5.11) is fulfilled. Suppose that $T : B[\rho_0] \to \mathcal{B}(X)$ satisfies condition (7.2.1) and the set*

$$\left\{(z - r)^2\big((z - r)^2 - r^2(I - T(0))\big)^{-1} : |z| = \alpha r, z \neq \alpha r\right\}$$

is *R*-bounded. Then there are a radius $0 < \rho \leq \rho_0$ and $\delta > 0$ such that for each $f \in l_p(\mathbb{Z}_+; X)$ with $\|f\|_p \leq \delta$ there is a unique $x = (x_n) \in MR_p$ with $\|x\|_{MR_p} \leq \rho$ satisfying

$$
\begin{cases}
\Delta_r^2 x_n - r^2(I - T(x_n))x_n = f_n, \; n \in \mathbb{Z}_+, \\[2mm]
x_0 = x_1 = 0.
\end{cases}
\tag{7.2.6}
$$

We obtain as an immediate consequence the following corollary.

Corollary 7.2.3. *Let H be a Hilbert space, $T(0) \in \mathcal{B}(H)$ be an analytic operator, and assume that (5.5.11) is fulfilled. Suppose that $T : B[\rho_0] \to \mathcal{B}(H)$ satisfies condition (7.2.1) and that*

$$
\sup_{|z| = \alpha r, \, z \neq \alpha r} \left\| (z - r)^2 R((z - r)^2, r^2(I - T(0))) \right\| < +\infty.
$$

Then, the conclusions of the Corollary 7.2.2 are true.

As an application, we consider the following examples.

Example 7.2.4. We consider the following maximal regularity space:

$$
MR_p := \{ y = (y_n) \in l_p(\mathbb{Z}_+; \mathbb{R}) : y_0 = y_1 = 0 \}.
$$

Then there are $\rho > 0$ and $\delta > 0$ such that for each $f \in l_p(\mathbb{Z}_+; \mathbb{R})$ with $\|f\|_p \leq \delta$ there is a unique $x = (x_n) \in MR_p$ satisfying $\|x\|_{MR_p} \leq \rho$ such that

$$
\begin{cases}
x_{n+2} - x_{n+1} + \dfrac{1}{4} \sin(x_n)x_n = f_n, \; n \in \mathbb{Z}_+, \\[3mm]
x_0 = x_1 = 0.
\end{cases}
$$

Example 7.2.5. We consider the spaces

$$
H := l_2(\mathbb{Z}_+ - \{0\}; \mathbb{R}) \quad \text{and} \quad MR_p := l_{p,r}^2(\mathbb{Z}_+; H) \cap l_p(\mathbb{Z}_+; H).
$$

Let $m : \mathbb{R} \to \mathbb{R}$ be a Lipschitz continuous function on bounded sets such that $m(0) = 0$. Then there are $\rho > 0$ and $\delta > 0$ such that for each $h \in l_p(\mathbb{Z}_+; H)$ with $\|h\|_{l_p(\mathbb{Z}_+;H)} \leq \delta$ there is a unique $x = (x_n) \in MR_p$ satisfying $\|x\|_{MR_p} \leq \rho$ such that

$$
\begin{cases}
x_{n+2}^l - 2r x_{n+1}^l + \dfrac{r^2}{l} m(x_n^l)x_n^l = h_n^l, \; n \in \mathbb{Z}_+, \, l \in \mathbb{Z}_+ - \{0\}, \\[3mm]
x_0^l = x_1^l = 0, \, l \in \mathbb{Z}_+ - \{0\}.
\end{cases}
\tag{7.2.7}
$$

Proof. For $\rho > 0$ denote by

$$B[\rho] := \{f \in H : \|f\|_2 \le \rho r^{-2}\}.$$

We define the function $T : B[\rho] \to \mathcal{B}(H)$ by

$$T(f)(g)(n) = m(f(n))A(g)(n), \quad f \in B[\rho], g \in H, n \in \mathbb{Z}_+ - \{0\},$$

where $A((g_n)_n) = (\frac{1}{n}g_n)_n$. To simplify the notation we write T as follows:

$$T(f) = m(f)A.$$

We note that[5] $A, T(f) \in \mathcal{B}(H), f \in B[\rho]$. Since $T(0) = 0$, the equation (7.2.2) associated to the operator T is

$$\begin{cases} x_{n+2}^l - 2rx_{n+1}^l = g_n^l, \; n \in \mathbb{Z}_+, l \in \mathbb{Z}_+ - \{0\}, \\ x_0^l = x_1^l = 0, \; l \in \mathbb{Z}_+ - \{0\}. \end{cases} \tag{7.2.8}$$

Observe that for $z \in \mathbb{T}_r^\alpha$, we have the estimate

$$\frac{|(z-r)^2|}{|z^2 - 2zr|} \le \frac{(|z|+r)^2}{|z|(|z|-2r)}$$

$$\le \frac{(\alpha+1)^2}{\alpha(\alpha-2)} = (\alpha+1)^2.$$

Hence

$$\sup_{z \in \mathbb{T}_r^\alpha} \frac{|(z-r)^2|}{|z^2 - 2zr|} \le (\alpha+1)^2.$$

By Corollary 5.6.4 the problem (7.2.8) is well posed.

---

[5]We have

$$\|Ag\|_2 \le \Big(\sum_{n=0}^{\infty} \frac{1}{n^2}\Big)^{1/2} \|g\|_2, \; g \in H,$$

and

$$\|T(f)\|_{\mathcal{B}(H)} \le L\|A\|_{\mathcal{B}(H)}\|f\|_2, \; f \in B[\rho].$$

Since m is a Lipschitz function on bounded sets, there exists $L > 0$ such that

$$|m(s) - m(t)| \leq L|s - t|, \quad \text{if } |s|, |t| \leq \rho r^{-2}.$$

If $f_1, f_2 \in B[\rho]$, then

$$|f_1(n)| \leq \rho r^{-2} \quad \text{and} \quad |f_2(n)| \leq \rho r^{-2}$$

for all $n \in \mathbb{Z}_+ - \{0\}$. Consequently, for $g \in H$, we obtain

$$\|(T(f_1) - T(f_2))g\|_2 \leq L\|A\|_{\mathcal{B}(H)}\|f_1 - f_2\|_2\|g\|_2.$$

In fact,

$$\|(T(f_1) - T(f_2))g\|_2 = \Big(\sum_{n=0}^{\infty} |m(f_1(n)) - m(f_2(n))|^2 |A(g)(n)|^2 \Big)^{1/2}$$

$$\leq L\Big(\sum_{n=0}^{\infty} |f_1(n) - f_2(n)|^2 |A(g)(n)|^2 \Big)^{1/2}$$

$$\leq L\|f_1 - f_2\|_2\|A(g)\|_2$$

$$\leq L\|A\|_{\mathcal{B}(H)}\|f_1 - f_2\|_2\|g\|_2.$$

Thus we have shown that

$$\|T(f_1) - T(f_2)\|_{\mathcal{B}(H)} \leq L\|A\|_{\mathcal{B}(H)}\|f_1 - f_2\|_2,$$

whenever $\|f_1\|_2 \leq \rho r^{-2}$, $\|f_2\|_2 \leq \rho r^{-2}$. Therefore (7.2.1) is fulfilled. Applying Corollary 7.2.3 we conclude the desired result.   □

Remark 7.2.6. It is easy to see that, repeating most parts of proof, the same type of result as in Theorem 7.2.1 holds for the following evolution problem:

$$\begin{cases} \Delta_r^2 x_n - r^2(I - T(x_n, \Delta_r x_n))x_n = G(n, x_n, \Delta_r x_n) + f_n, \ n \in \mathbb{Z}_+, \\ \\ x_0 = x_1 = 0, \end{cases} \tag{7.2.9}$$

where $T : B[\rho_0] \times B[\rho_0] \to \mathcal{B}(X)$ is a function such that

$$\|T(x_1, y_1) - T(x_2, y_2)\|_{\mathcal{B}(X)}$$
$$\leq L\Big(\|(I - T(0,0))(x_1 - x_2)\|_X + \|(I - T(0,0))(y_1 - y_2)\|_X \Big).$$

7.3   An *R*-Boundedness Criterion

On an arbitrary Banach space, in general, it is a difficult task to verify
R-boundedness. Here, we are interested in investigating when the set

$$\mathcal{M} := \left\{ (z-r)^2 \left((z-r)^2 - r^2 (I-T) \right)^{-1} \ : \ |z| = \alpha r, \ z \neq \alpha r \right\} \tag{7.3.1}$$

defined in (ii) of Theorem 5.6.3 (see also Theorem 6.3.2) is *R*-bounded. The following theorem has practical importance because it provides a sufficient condition
to assure the *R*-boundedness of the abovementioned set.

Theorem 7.3.1. *Assume that (5.5.11) is fulfilled and suppose that the family of sine
operators* $\{\mathcal{S}(n)\}_{n \in \mathbb{Z}_+}$ *(see Sect. 1.4) satisfies the following condition: there is a
positive constant C such that*

$$\sum_{n=1}^{\infty} ||r^{n-1} \mathcal{S}(n) x||_X \leq C ||x||_X, \quad \text{for all } x \in X. \tag{7.3.2}$$

Then the sets

$$\mathcal{T}_S = \left\{ \mathcal{F}[r^{\cdot - 1} \mathcal{S}](z) : |z| = \alpha r, \ z \neq \alpha r \right\}$$

and

$$\mathcal{T}_C = \left\{ \mathcal{F}[r^{\cdot} \mathcal{C}](z) : |z| = \alpha r, \ z \neq \alpha r \right\}$$

are R-bounded (here $\mathcal{C} := \{\mathcal{C}(n)\}_{n \in \mathbb{Z}_+}$ *denotes the family of cosine operators).
Moreover,*

$$R(\mathcal{T}_S) \leq 2C \quad \text{and} \quad R(\mathcal{T}_C) \leq 2(1+r)C.$$

In particular, if $T \in \mathcal{B}(X)$ *is an analytic operator, then* $\mathcal{M}$ *is R-bounded.*

Proof. Denote by r_n the *n*th Rademacher function in $[0, 1]$, i.e.,

$$r_n(t) = sgn(\sin(2^n \pi t)).$$

For $x \in X$, we denote by $r_n \otimes x$ the vector-valued function

$$t \longrightarrow r_n(t) x.$$

Take $z_k = \alpha r \exp(i \theta_k)$, $k = 1, \ldots, n$, $h_k(j) = z_k^{-j}$ and $x_1, \ldots, x_n \in X$.

We have[6]

$$\left(\sum_{n=1}^{m} r_n \otimes \mathcal{F}[r^{\bullet-1}\mathcal{S}](z_n)x_n \right)(t) = \sum_{n=1}^{m} \left(r_n \otimes \mathcal{F}[r^{\bullet-1}\mathcal{S}](z_n)x_n \right)(t)$$

$$= \sum_{n=1}^{m} \sum_{j=1}^{\infty} r_n(t)h_n(j)r^{j-1}\mathcal{S}(j)x_n$$

$$= \sum_{j=1}^{\infty} \sum_{n=1}^{m} r_n(t)h_n(j)r^{j-1}\mathcal{S}(j)x_n$$

$$= \sum_{j=1}^{\infty} \sum_{n=1}^{m} \left(r_n \otimes h_n(j)r^{j-1}\mathcal{S}(j)x_n \right)(t)$$

$$= \left(\sum_{j=1}^{\infty} \sum_{n=1}^{m} r_n \otimes h_n(j)r^{j-1}\mathcal{S}(j)x_n \right)(t).$$

$$(7.3.3)$$

Using identity (7.3.3) and taking advantage of the Kahane contraction principle[7] (see Proposition 2.5.6), we get

$$\left\| \sum_{n=1}^{m} r_n \otimes \mathcal{F}[r^{\bullet-1}\mathcal{S}](z_n)x_n \right\|_{L^1([0,1];X)}$$

$$= \left\| \sum_{j=1}^{\infty} \sum_{n=1}^{m} r_n \otimes h_n(j)r^{j-1}\mathcal{S}(j)x_n \right\|_{L^1([0,1];X)}$$

$$\leq \sum_{j=1}^{\infty} \left\| \sum_{n=1}^{m} r_n \otimes h_n(j)r^{j-1}\mathcal{S}(j)x_n \right\|_{L^1([0,1];X)}$$

$$\leq 2 \sum_{j=1}^{\infty} \left\| \sum_{n=1}^{m} r_n \otimes r^{j-1}\mathcal{S}(j)x_n \right\|_{L^1([0,1];X)}$$

---

[6] To prove Theorem 7.3.1 we follow Kunstmann and Weis' argument of proof [127, Corollary 2.17].

[7] Note that $|h_n(j)| = \left| \dfrac{1}{z_n} \right|^j = \left(\dfrac{1}{|z_n|} \right)^j = \left(\dfrac{1}{\alpha r} \right)^j \leq 1.$

$$= 2 \sum_{j=1}^{\infty} \int_0^1 \Big\| \sum_{n=1}^{m} r_n(t) r^{j-1} S(j) x_n \Big\|_X dt$$

$$\leq 2 \int_0^1 \Big(\sum_{j=1}^{\infty} \big\| r^{j-1} S(j) \big\| \sum_{n=1}^{m} r_n(t) x_n \Big\|_X \Big) dt$$

$$\leq 2C \int_0^1 \Big\| \sum_{n=1}^{m} r_n(t) x_n \Big\|_X dt$$

$$= 2C \Big\| \sum_{n=1}^{m} r_n \otimes x_n \Big\|_{L^1([0,1];X)},$$

we conclude that

$$\mathcal{T}_S \text{ is } R\text{-bounded and } R(\mathcal{T}_S) \leq 2C.$$

To prove that $\mathcal{T}_C$ is R-bounded, we observe that $C(n) = \Delta S(n)$, $n \in \mathbb{Z}_+$ (see Proposition 1.4.4); hence

$$\sum_{n=1}^{\infty} \|r^n C(n) x\|_X = \sum_{n=1}^{\infty} \Big\| r^n \big(S(n+1) - S(n) \big) x \Big\|_X$$

$$\leq \sum_{n=1}^{\infty} \|r^{n-1} S(n) x\|_X$$

$$+ r \sum_{n=1}^{\infty} \|r^{n-1} S(n) x\|_X$$

$$\leq C \|x\|_X + rC \|x\|_X$$

$$= (1+r) C \|x\|_X. \tag{7.3.4}$$

Next using (7.3.4) we have the following estimate:

$$\left\| \sum_{n=1}^{m} r_n \otimes \mathcal{F}[r^{\bullet}C](z_n)x_n \right\|_{L^1([0,1];X)}$$

$$\leq \sum_{j=0}^{\infty} \left\| \sum_{n=1}^{m} r_n \otimes h_n(j)r^j C(j)x_n \right\|_{L^1([0,1];X)}$$

$$\leq 2 \sum_{j=0}^{\infty} \left\| \sum_{n=1}^{m} r_n \otimes r^j C(j)x_n \right\|_{L^1([0,1];X)}$$

$$\leq 2(1+r)C \left\| \sum_{n=1}^{m} r_n \otimes x_n \right\|_{L^1([0,1];X)}.$$

This last estimate guarantees that

$$\mathcal{T}_C \text{ is } R\text{-bounded and } R(\mathcal{T}_C) \leq 2(1+r)C.$$

Now if $T \in \mathcal{B}(X)$ is an analytic operator, taking into account the R-boundedness, basic operations (see Proposition 2.2.5), Proposition 5.5.5, and the identity

$$(z-r)^2 R((z-r)^2, r^2(I-T)) = I + r^2(I-T)R((z-r)^2, r^2(I-T))$$
$$= I + z^{-1}r^2(I-T)\mathcal{F}[r^{\bullet-1}S](z),$$

it follows that the set $\mathcal{M}$ is R-bounded. This finishes the proof. □

Remark 7.3.2. In the following case condition (7.3.2) is fulfilled:

(i) $T = I$ and $1/(1 + \sqrt{2}) < r < 1$. In fact, we have

$$\sum_{n=1}^{\infty} \|r^{n-1}S(n)x\|_X \leq \left(\frac{1}{r} \sum_{n=1}^{\infty} nr^n \right) \|x\|_X.$$

(ii) $T = i\epsilon I$, with $\epsilon > 0$ small enough and $1/(1 + \sqrt{2}) < r < 1/(1 + \sqrt[4]{1+\epsilon^2})$. In fact, putting $\rho = \sqrt{1+\epsilon^2}$ and $\alpha = \arctan(\epsilon)$, it follows from (1.4.6) that

$$\sqrt{\rho}e^{-i\alpha/2}S(n) = \frac{1}{2}\left((1 + \sqrt{\rho}e^{-i\alpha/2})^n - (1 - \sqrt{\rho}e^{-i\alpha/2})^n \right)I.$$

Hence

$$S(n) = \sum_{j=1}^{n-1} (1 + \sqrt{\rho}e^{-i\alpha/2})^{n-1-j}(1 - \sqrt{\rho}e^{-i\alpha/2})^j I.$$

Consequently,

$$\sum_{n=1}^{\infty} ||r^{n-1} S(n) x||_X \le \left(\frac{1}{r(1 + \sqrt{\rho})} \sum_{n=1}^{\infty} n r^n (1 + \sqrt{\rho})^n\right) ||x||_X.$$

Since $r < 1/(1 + \sqrt[4]{1 + \epsilon^2})$, we find that the series $\sum_{n=1}^{\infty} n r^n (1 + \sqrt{\rho})^n$ converges. We observe that since $\epsilon > 0$ is small enough T is analytic.

Corollary 7.3.3. *Let X be a UMD space. If f is defined and satisfies a Lipschitz condition in X and let $q = (q_n)$ be a summable complex sequence and $r \in]1/(1 + \sqrt{2}), \sqrt{2}/(\sqrt{2} + \sqrt[4]{5})[$, then the following problem:*

$$\begin{cases} x_{n+2} - 2r x_{n+1} + \dfrac{i}{2} r^2 x_n = q_n f(x_n), \ n \in \mathbb{Z}_+, \\[2mm] x_0 = x_1 = 0, \end{cases} \qquad (7.3.5)$$

has a unique solution $x = (x_n)$ in $l_p(\mathbb{Z}_+; X)$.

Proof. Taking $\epsilon = \frac{1}{2}$ and $T = \frac{i}{2} I$ in Remark 7.3.2(ii) we find that T is analytic and the set $\mathcal{M}$ in (7.3.1) is R-bounded. Since f is Lipschitz and $q = (q_n)$ is summable, it follows from Theorem 6.3.2 that (7.3.5) has a unique solution in $l_p(\mathbb{Z}_+; X)$. □

As a consequence of Theorems 7.3.1 and 5.6.3, we have the following result.

Theorem 7.3.4. *Let X be a UMD space and let $T \in \mathcal{B}(X)$ be an analytic operator. In addition, assume that (5.5.11) and (7.3.2) are fulfilled. Then problem (5.5.8) is well posed.*

7.4 Stability

The growth of the theory of discrete evolution equations has been strongly promoted by the advanced technology in scientific computation and the large number of applications in models in biology, engineering, and other physical sciences. It is the stability of solutions of these models that is especially important to many researchers.

The stability of a discrete process is the capacity of the process to resist a priori unknown small influences. A process is said to be stable if such disturbances do not change it drastically.

The following results provide a new criterion to verify the stability of discrete semilinear systems. Note that the characterization of maximal regularity is the key to give conditions based only on the data of a given system.[8]

---

[8]Of course, all results of this section hold in the finite dimensional case.

Theorem 7.4.1. *Let X be a UMD space. Assume that condition (A_1) in Sect. 4.1 holds and suppose $T \in \mathcal{B}(X)$ is analytic and $1 \in \rho(T)$. Then the system (4.1.2) is stable, that is, the solution (x_n) of (4.1.2) is such that $x_n \to 0$ as $n \to \infty$.*

Proof. It is assumed that T is analytic (which implies that the spectrum is contained in the unit disc and the point 1; see Theorem 1.3.9) and that 1 is not in the spectrum, then in view of Proposition 2.2.5(h), the set

$$\{(\lambda - 1)R(\lambda, T) : |\lambda| = 1, \ \lambda \neq 1\}, \tag{7.4.1}$$

is R-bounded, because $(\lambda - 1)R(\lambda, T)$ is an analytic function in a neighborhood of the circle. The power boundedness assumption of operator T follows from Proposition 2.5.10. By Corollary 4.1.2, there exists a unique solution (x_n) of (4.1.2) such that $(\Delta x_n) \in l_p(\mathbb{Z}_+, X)$. Then

$$\Delta x_n \to 0, \ \text{as } n \to \infty.$$

Next, observe that condition (A_1) implies

$$\|f(n, x_n)\|_X \leq \|f(n, x_n) - f(n, 0)\|_X + \|f(n, 0)\|_X$$

$$\leq \alpha_n \|x_n\|_X + \|f(n, 0)\|_X. \tag{7.4.2}$$

Since $(x_n) \in l_\infty(\mathbb{Z}_+, X)$, $(f(n, 0)) \in l_1(\mathbb{Z}_+, X)$, and $(\alpha_n) \in l_1(\mathbb{Z}_+)$, we obtain that $f(n, x_n) \to 0$ as $n \to \infty$. Then, the result follows from the following identity:

$$x_n = (I - T)^{-1}[f(n, x_n) - \Delta X_n].$$

$\square$

From the point of view of applications, the following corollary provides practical conditions for stability.

Corollary 7.4.2. *Let H be a Hilbert space. Let $T \in \mathcal{B}(H)$ be such that $\|T\| < 1$. Suppose that the following conditions hold:*

(i) *The function $f : \mathbb{Z}_+ \times H \longrightarrow H$ satisfies: for all $z, w \in H$ and $n \in \mathbb{Z}_+$, we have*

$$\|f(n, z) - f(n, w)\|_H \leq \alpha_n \|z - w\|_H,$$

where $\alpha := (\alpha_n) \in l_1(\mathbb{Z}_+)$.
(ii) *$f(\cdot, 0) \in l_1(\mathbb{Z}_+; H)$.*

Then the solution of the problem (4.1.2) is stable.

Proof. First we note that each Hilbert space is UMD, and then the concept of R-boundedness and boundedness coincide; see Proposition 2.2.5(d). Since $||T|| < 1$, we find that T is power bounded and analytic and $1 \in \rho(T)$. Furthermore, for $|\lambda| = 1$, $\lambda \neq 1$, the inequality

$$||(\lambda - 1)R(\lambda, T)|| = ||(\lambda - 1) \sum_{n=0}^{\infty} \left(\frac{T}{\lambda}\right)^n|| \leq \frac{|\lambda - 1||\lambda|}{1 - ||T||} < \frac{2}{1 - ||T||},$$

shows that the set (7.4.1) is bounded. $\square$

As an application, we consider a semilinear discrete control system of the form

$$x_{n+1} = Ax_n + Bu_n + F(x_n, u_n), \quad n \in \mathbb{Z}_+, \tag{7.4.3}$$

where A and B are constant matrices, F is a nonlinear function, and u_n is a control input.

The system (7.4.3) was considered in [157]. Based on the state space quantization method used in [64, 153], the authors established sufficient conditions for the global stabilizability of the semilinear discrete-time system under an appropriate growth condition on the nonlinear perturbation. In contrast with this approach, applying Corollary 7.4.2 to $f(n, z) = Bu_n + F(z, u_n)$ we directly obtain the following remarkable result.

Proposition 7.4.3. *Suppose that the following conditions hold:*

(i) *The function $F : \mathbb{R}^N \times \mathbb{R}^N \longrightarrow \mathbb{R}^N$ satisfies: for all $z, w \in \mathbb{R}^N$ and $n \in \mathbb{Z}_+$, we have $||F(z, u_n) - F(w, u_n)|| \leq \alpha_n ||z - w||$, where $\alpha := (\alpha_n) \in l_1(\mathbb{Z}_+)$.*
(ii) *$Bu_n + F(0, u_n) \in l_1(\mathbb{Z}_+)$.*
(iii) *$||A|| < 1$.*

Then system (7.4.3) is stable.

Example 7.4.4. Let α, β, c, d be positive real numbers. The following semilinear discrete control system is considered in [157]:

$$x_{n+1} = \alpha x_n + \beta u_n + F(x_n, u_n), \quad n \in \mathbb{Z}_+, \tag{7.4.4}$$

where

$$F(x_n, u_n) = cx_n \sin u_n + d u_n \cos^2 x_n, \quad n \in \mathbb{Z}_+.$$

Here we have $\alpha_n = (c + 2d)|u_n|$. As a consequence of Proposition 7.4.3, we find that for all $0 < \alpha < 1$ and each control input u_n in $l_1(\mathbb{Z}_+)$, the solution of system (7.4.4) converges to zero as $n \to \infty$.

Concerning second-order systems, we have the following result.

Theorem 7.4.5. *Let X be a UMD space. Assume that condition (A_2) in Sect. 6.1 holds and suppose $T \in \mathcal{B}(X)$ is analytic and $1 \in \rho(T)$. Then the system (6.1.2) is stable.*

Proof. In view of Proposition 2.2.5(h), we note that the set

$$\left\{ (\lambda - 1)^2 R((\lambda - 1)^2, I - T) : |\lambda| = 1, \lambda \neq 1 \right\} \tag{7.4.5}$$

is R-bounded. Hence Theorem 7.4.5 is a consequence of Corollary 6.1.2 and the fact that $\Delta^2 x_n \to 0$ and $f(n, x_n, \Delta x_n) \to 0$ as $n \to \infty$, where (x_n) is the solution of (6.1.2). ☐

We have the following immediate corollary.

Corollary 7.4.6. *Let H be a Hilbert space. Let $T \in \mathcal{B}(H)$ be such that $||T|| < 1$. Suppose that Assumption (A_2) holds in H. Then the system (6.1.2) is stable.*

The following result provides a new criterion to verify the stability of semilinear discrete harmonic oscillator equation.

Theorem 7.4.7. *Let X be a UMD space. Assume that condition (A_3) in Sect. 6.2 holds. In addition, suppose $T \in \mathcal{B}(X)$ is an analytic operator with $1 \in \rho(T)$. Then, (6.2.1) is stable.*

Proof. Since $\sigma(T) \subseteq \mathbb{D} \cup \{1\}$ and 1 is not in $\sigma(T)$, the set (6.2.27) is R-bounded. By Corollary 6.2.3, there exists a unique solution $x = (x_n)$ of (6.2.1) such that $(\Delta^2 x_n) \in l_p(\mathbb{Z}_+; X)$. Then $\Delta^2 x_n \to 0$ as $n \to \infty$. Next, we observe that condition (A_3) and the estimate (6.2.6) imply

$$||f(n, x_n, \Delta x_n)||_X \leq \alpha_n ||(x_n, \Delta x_n)||_{X \times X} + ||f(n, 0, 0)||_X$$

$$\leq \frac{\alpha_n}{2}(n^2 + n)||\Delta^2 x||_p + ||f(n, 0, 0)||_X$$

$$\leq \alpha_n 2^{2n} ||\Delta^2 x||_p + ||f(n, 0, 0)||_X.$$

Since $(f(\cdot, 0, 0)) \in l_1(\mathbb{Z}_+, X)$ and $\alpha_n 2^{2n} \to 0$, as $n \to \infty$, we find that $f(n, x_n, \Delta x_n) \to 0$ as $n \to \infty$. The result now follows from the hypothesis that $1 \in \rho(T)$ and (6.2.1). ☐

Theorem 7.4.8. *Let X be a UMD space. Assume that (5.5.11) and (A_4) in Sect. 6.3 hold. In addition, suppose that $T \in \mathcal{B}(X)$ is an analytic operator with $1 \in \rho(T)$ and the set $\mathcal{M}$ in (7.3.1) is R-bounded. Then, the system (6.3.1) is stable.*

Proof. Since $((A_4))$ is fulfilled and the set $\mathcal{M}$ is R-bounded. By Theorem 6.3.2 there is a unique solution (x_n) of (6.3.1) such that $(\Delta_r^2 x_n)$ belongs to $l_p(\mathbb{Z}_+; X)$. Then

$$\Delta_r^2 x_n \to 0 \text{ as } n \to \infty. \tag{7.4.6}$$

Moreover, one has the following a priori estimate for the solution (x_n):

$$\sup_{n\in\mathbb{Z}_+}\left(\frac{1}{\gamma(r,n)}\left(||x_n||_X+||\Delta_r x_n||_X\right)\right)\le ||G(\cdot,0,0)||_1||\rho\cdot\gamma(r,\cdot)||_1^{-1}e^{2A(r)},$$

$$(7.4.7)$$

where $A(r)$ is given in (6.3.2). Using Assumption (A$_4$) and (7.4.7) we obtain the following estimate:

$$||G(n,x_n,\Delta_r x_n)||\le ||G(n,x_n,\Delta_r x_n)-G(n,0,0)||+||G(n,0,0)||$$

$$\le \rho_n\left(||x_n||_X+||\Delta_r x_n||_X\right)+||G(n,0,0)||$$

$$\le \rho_n\gamma(r,n)\sup_{n\in\mathbb{Z}_+}\left(\frac{1}{\gamma(r,n)}\left(||x_n||_X+||\Delta_r x_n||_X\right)\right)$$

$$+||G(n,0,0)||$$

$$\le \rho_n\gamma(r,n)||G(\cdot,0,0)||_1||\rho\cdot\gamma(r,\cdot)||_1^{-1}e^{2A(r)}$$

$$+||G(n,0,0)||.$$

Taking into account that $\lim_{n\to\infty}\rho_n\gamma(r,n)=0$ and $\lim_{n\to\infty}||G(n,0,0)||=0$, we obtain

$$\lim_{n\to\infty}||G(n,x_n,\Delta_r x_n)||=0.\qquad(7.4.8)$$

We observe that since (x_n) is a solution of (6.3.1) and $1\in\rho(T)$

$$x_n=\frac{1}{r^2}(I-T)^{-1}\left(\Delta_r^2 x_n-G(n,x_n,\Delta_r x_n)\right).$$

From (7.4.6) and (7.4.8) we conclude that

$$x_n\to 0\quad\text{as}\quad n\to\infty.$$

$$\square$$

The following corollary is a consequence of the preceding theorem and Proposition 2.2.6(a).

Corollary 7.4.9. *Let X be a UMD space. Assume that (5.5.11) and (A$_4$) hold. Let T be in $\mathcal{B}(X)$ such that $||T||<1$ and the set $\{(I-T)^n:n\in\mathbb{Z}_+\}$ be R-bounded. Then the system (6.3.1) is stable.*

Proof. For $z\in\mathbb{C}$ with $|z|=\alpha r$, we have $|z-r|>\sqrt{2}r$. Since $||T||<1$, we get

$$\left\| \frac{r^2}{(z-r)^2}(I-T) \right\| = \frac{r^2}{|z-r|^2} \|I-T\|$$

$$\leq \frac{1}{2}(1 + \|T\|) < 1.$$

Hence

$$\left(I - \frac{r^2}{(z-r)^2}(I-T) \right)^{-1} = \sum_{n=0}^{\infty} \frac{r^{2n}}{(z-r)^{2n}}(I-T)^n$$

is well defined. Since the set

$$\left\{ \left(\frac{r^2}{(z-r)^2} \right)^n : n \in \mathbb{Z}_+ \right\} \subseteq \mathbb{C}$$

is bounded,[9] by Proposition 2.2.5(g) the set

$$\left\{ \frac{r^{2n}}{(z-r)^{2n}}I : n \in \mathbb{Z}_+ \right\} \text{ is } R\text{-bounded.}$$

From the fact that the set $\{(I-T)^n : n \in \mathbb{Z}_+\}$ is R-bounded and using Proposition 2.2.5(f) we find that the set

$$\left\{ \frac{r^{2n}}{(z-r)^{2n}}(I-T)^n : n \in \mathbb{Z}_+ \right\} \text{ is } R\text{-bounded.}$$

Hence by Proposition 2.2.6(a) we get that $\mathcal{M}$ is R-bounded. □

Corollary 7.4.10. *Assume that (5.5.11) and* (A$_4$) *in Sect. 6.3 hold in a Hilbert space H. Let T be in $\mathcal{B}(H)$ such that $\|T\| < 1$. Then, the system (6.3.1) is stable.*

Proof. For $|z| = \alpha r$ with $z \neq \alpha r$. Keeping in mind (5.5.11), we have the following estimates:

$$\left\| \frac{r^2}{z(z-2r)}T \right\| \leq \frac{r^2}{|z|(|z|-2r)}\|T\|$$

$$\leq \frac{1}{(\sqrt{2}+1)(\sqrt{2}-1)}\|T\| = \|T\| < 1,$$

---

[9] We note that $\frac{r^2}{|z-r|^2} \leq \frac{1}{2}$ for all $z \in \mathbb{C}$ with $|z| = \alpha r$.

and

$$\|(z-r)^2 R((z-r)^2, r^2(I-T))\| = \left\| \frac{(z-r)^2}{z(z-2r)} \left(I - \frac{r^2}{z^2 - 2zr} T \right)^{-1} \right\|$$

$$= \frac{|z-r|^2}{|z||z-2r|} \left\| \sum_{n=0}^{\infty} \left(\frac{r^2}{z^2 - 2zr} T \right)^n \right\|$$

$$\leq \frac{|z-r|^2}{|z||z-2r| - r^2 \|T\|}$$

$$\leq \frac{(|z|+r)^2}{r^2(1 - \|T\|)}$$

$$\leq \frac{(\sqrt{2}+2)^2}{r^2(1 - \|T\|)}$$

$$\leq \frac{16}{1 - \|T\|},$$

which are responsible for the fact that the set

$$\{(z-r)^2((z-r)^2 - r^2(I-T))^{-1} \; : \; |z| = \alpha r, z \neq \alpha r\}$$

in (7.3.1) is bounded. The proof now follows from Theorem 7.4.8.   □

Remark 7.4.11. Suppose that $T \in \mathcal{B}(X)$ is an analytic operator with $1 \in \rho(-T)$, then the set $\{(z-r)^2((z-r)^2 - r^2(I-T))^{-1} \; : \; |z| = \alpha r, z \neq \alpha r\}$ is R-bounded. In fact, since T is analytic, by Proposition 5.5.4 , $(z-r)^2 \in \rho(r^2(I-T))$ whenever $|z| = \alpha r, z \neq \alpha r$. If it is assumed that $2r^2$ is not in the spectrum of $r^2(I-T)$, then in view of Proposition 2.2.5(h), we conclude our assertion.

Example 7.4.12. Let X be a UMD space and assume that (5.5.11) is fulfilled. Let f be defined and satisfy a Lipschitz condition in X and let $q = (q_n)$ be a complex sequence such that $\sum_{n=0}^{\infty} |q_n| \gamma(r, n) < +\infty$. Consider the semilinear evolution problem in X:

$$\begin{cases} x_{n+2} - 2r x_{n+1} = q_n f(x_n), \ n \in \mathbb{Z}_+, \\ \\ x_0 = x_1 = 0. \end{cases} \qquad (7.4.9)$$

Then, by Theorem 7.4.8, the unique solution $x = (x_n)$ of (7.4.9) is such that $x_n \to 0$ as $n \to \infty$. Indeed, let z be a complex number such that $|z| = \alpha r$, then taking into account that $\alpha(\alpha - 2) = 1$ [see (5.5.11)] and $|z| > 2r$ we obtain

$$\left| \frac{(z-r)^2}{(z-r)^2 - r^2} \right| = \frac{|z-r|^2}{|z||z-2r|}$$

$$\leq \frac{(|z|+r)^2}{|z|(|z|-2r)}$$

$$= \frac{(\alpha+1)^2 r^2}{\alpha(\alpha-2)r^2} = (\alpha+1)^2.$$

Hence the set

$$\left\{ \frac{(z-r)^2}{(z-r)^2 - r^2} : |z| = \alpha r \right\}$$

is bounded; now by Proposition 2.2.5(g) the set

$$\left\{ \frac{(z-r)^2}{(z-r)^2 - r^2} I : |z| = \alpha r \right\}$$

is R-bounded.

7.5 Boundedness

The main result of this section ensures the existence and uniqueness of weighted bounded solutions under quite general hypotheses. Before discussing the result, we need to introduce the following technical requirements.

Assumption $(A_1)^{**}$: Let $(c(r,n))_n$ be a positive sequence which is nondecreasing with respect to the second variable. Suppose that the following conditions hold:

(i)** The function $G : \mathbb{Z}_+ \times X \times X \longrightarrow X$ satisfies a Lipschitz condition in $X \times X$, i.e., for all $z, w \in X \times X$ and $n \in \mathbb{Z}_+$, we have

$$\|G(n,z) - G(n,w)\|_X \leq \rho_n \|z - w\|_{X \times X},$$

where $\rho := (\rho_n)$ is a positive sequence such that

$$\sum_{n=0}^{\infty} \rho_n c(r, n+1) < +\infty.$$

(ii)** $G(\cdot, 0, 0) \in l_1(\mathbb{Z}_+; X)$.

With the above notations, we have the following result.

Theorem 7.5.1. *Let X be a Banach space and $T \in \mathcal{B}(X)$. Assume that the following conditions hold:*

(a)** *Assumption $(A_1)^{**}$ is fulfilled.*

(b)** $M := \sup_{n \in \mathbb{Z}_+} \left(\dfrac{||r^{n-1}S(n)||}{c(r, n)} + \dfrac{||r^n C(n)||}{c(r, n+1)} \right) < +\infty.$

Then, there is a positive constant $m \in \mathbb{Z}_+$ and a unique weighted bounded [see (7.5.5)] solution $y = (y_n)$ of (6.3.1) for $n \geq m$ such that $y_1 = \cdots = y_m = 0$. Moreover, one has the following a priori estimate:

$$\sup_{n \in \mathbb{Z}_+} \left(\frac{||y_n||_X}{c(r, n)} + \frac{||\Delta_r y_n||_X}{c(r, n+1)} \right) \leq 2M ||G(\cdot, 0, 0)||_1 e^{2M ||c(r, \cdot+1)\rho \cdot ||_1}. \quad (7.5.1)$$

Remark 7.5.2.

(i) Let $\Theta(r)$ be the constant in (5.5.15) and let $\gamma(r, n)$ be the sequence given by (5.5.16). Under the assumptions of Theorem 5.6.3, we suppose that the problem (5.5.8) is well posed in the sense of Definition 5.6.1. Then by Corollary 5.6.5 we get

$$\frac{||r^{n-1}S(n)||}{\gamma(r, n+1)} \leq (1 + ||\mathcal{K}^r||_{\mathcal{B}(l_p(\mathbb{Z}_+; X))})\Theta(r).$$

We note that the sequence $(\gamma(r, n))$ is nondecreasing with respect to the second variable, so using Corollary 5.6.5 again, we get

$$\frac{||r^n C(n)||}{\gamma(r, n+2)} \leq \frac{\gamma(r, n+1)}{\gamma(r, n+2)} \cdot \frac{||r^n C(n)||}{\gamma(r, n+1)}$$

$$\leq \frac{||r^n C(n)||}{\gamma(r, n+1)} \leq (1 + ||\mathcal{K}^r||_{\mathcal{B}(l_p(\mathbb{Z}_+; X))})\Theta(r).$$

Hence condition (b)** is fulfilled with

$$c(r, n) := \gamma(r, n+1) \text{ and } M := 2(1 + ||\mathcal{K}^r||_{\mathcal{B}(l_p(\mathbb{Z}_+; X))})\Theta(r).$$

(ii) We say that (5.5.8) has discrete maximal l_∞-regularity if $\mathcal{K}_\infty^r f := (I - T)r^{\cdot+1}S * f$ defines a linear bounded operator $\mathcal{K}_\infty^r \in \mathcal{B}(l_\infty(\mathbb{Z}_+; X))$. As a consequence of this definition, if (5.5.8) has discrete maximal l_∞-regularity, then for each $f = (f_n) \in l_\infty(\mathbb{Z}_+; X)$, we have $(\Delta_r^2 x_n) \in l_\infty(\mathbb{Z}_+; X)$, where (x_n) is the solution of (5.5.8).[10] We also note that if (5.5.8) has a discrete maximal l_∞-regularity, the condition (b)** of Theorem 7.5.1 holds. Now we define the following sequence space:

$$l_{\infty,r}^2(\mathbb{Z}_+; X) := \left\{ y = (y_n) : y_0 = y_1 = 0, (\Delta_r^2 y_n) \in l_\infty(\mathbb{Z}_+; X) \right\}.$$

We will next derive the condition (b)** from the following estimate:

Proposition 7.5.3. *For each* $y \in l_{\infty,r}^2(\mathbb{Z}_+; X)$ *and* $n \in \mathbb{Z}_+$, *we have the following a priori estimate:*

$$\frac{\|y_n\|_X}{\gamma(r, n)} + \frac{\|\Delta_r y_n\|_X}{\gamma(r, n + 1)} \le \Gamma(r)\|\Delta_r^2 y.\|_\infty, \tag{7.5.2}$$

where

$$\Gamma(r) = \begin{cases} \dfrac{2 - r}{(1 - r)^2}, & \text{for } r < 1, \\[3mm] \dfrac{r}{(r - 1)^2}, & \text{for } r > 1, \\[3mm] 1, & \text{for } r = 1. \end{cases} \tag{7.5.3}$$

Indeed, let $x \in X$ be an arbitrary vector and define

$$h_n = \begin{cases} x, & n = 0, \\[2mm] 0, & n \ne 0. \end{cases}$$

We consider the problem

$$\begin{cases} \Delta_r^2 y_n - r^2(I - T)y_n = h_n, & n \in \mathbb{Z}_+, \\[2mm] y_0 = 0, \ y_1 = 0. \end{cases} \tag{7.5.4}$$

---

[10]We observe that

$$\Delta_r^2 x_n = (\mathcal{K}_\infty^r f)_{n-1} + f_n.$$

Hence

$$\|\Delta_r^2 x\|_\infty \le (1 + \|\mathcal{K}_\infty^r\|_{B(l_\infty(\mathbb{Z}_+;X))})\|f\|_\infty.$$

Since $h = (h_n) \in l_\infty(\mathbb{Z}_+; X)$ and (7.5.4) has discrete maximal l_∞-regularity, we find that the solution (y_n) of (7.5.4) is such that $(\Delta_r^2 y_n)$ belongs to $l_\infty(\mathbb{Z}_+; X)$. Therefore

$$y \in l_{\infty,r}^2(\mathbb{Z}_+; X)$$

and

$$||\Delta_r^2 y||_\infty \leq (1 + ||\mathcal{K}_\infty^r||_{\mathcal{B}(l_\infty(\mathbb{Z}_+; X))})||x||_X.$$

We note by Corollary 5.5.2 the following representation for the solution (y_n) of (7.5.4):

$$y_{n+1} = r^{n-1} S(n)x \text{ and } \Delta_r y_{n+1} = r^n C(n)x.$$

From Proposition 7.5.3 we obtain

$$\frac{||r^{n-1}S(n)x||_X}{\gamma(r, n+1)} + \frac{||r^n C(n)x||_X}{\gamma(r, n+2)}$$

$$\leq \Gamma(r)(1 + ||\mathcal{K}_\infty^r||_{\mathcal{B}(l_\infty(\mathbb{Z}_+; X))})||x||_X,$$

Taking

$$c(r, n) := \gamma(r, n+1), \text{ and } M := 2(1 + ||\mathcal{K}_\infty^r||_{\mathcal{B}(l_\infty(\mathbb{Z}_+; X))})\Gamma(r),$$

the condition (b)** follows.

To finish the discussion of Remark 7.5.2 we prove the Proposition 7.5.3.

Proof of Proposition 7.5.3. We divide the proof into several cases[11]

---

[11]

(i) For $r < 1$: $\sum_{i=0}^{n-1} \sum_{k=i}^{n-2} r^k \leq \sum_{i=0}^{n-1} \frac{r^i}{1-r} \leq \frac{1}{(1-r)^2}$.

(ii) For $r > 1$: $\sum_{i=0}^{n-1} \sum_{k=i}^{n-2} r^k \leq \sum_{i=0}^{n-1} \frac{r^n}{r(r-1)} \leq \frac{nr^n}{(r-1)^2}$.

(iii) For $r = 1$: $\sum_{i=0}^{n-1} \sum_{k=i}^{n-2} 1 = \frac{n(n-1)}{2}$.

(a) The case $r < 1$. We have the following estimates[12]

$$\|y_n\|_X = \left\| \sum_{i=0}^{n-1} \sum_{j=0}^{n-2-i} r^{i+j} \Delta_r^2 y_{n-2-i-j} \right\|_X$$

$$\leq \left(\sum_{i=0}^{n-1} \sum_{k=i}^{n-2} r^k \right) \|\Delta_r^2 y.\|_\infty$$

$$\leq \frac{1}{(1-r)^2} \|\Delta_r^2 y.\|_\infty.$$

and

$$\|\Delta_r y_n\|_X = \left\| \sum_{i=0}^{n-1} r^i \Delta_r^2 y_{n-1-i} \right\|_X$$

$$\leq \left(\sum_{i=0}^{n-1} r^i \right) \|\Delta_r^2 y.\|_\infty$$

$$\leq \frac{1}{1-r} \|\Delta_r^2 y.\|_\infty.$$

(b) The case $r > 1$. We have the following estimates:

$$\|y_n\|_X \leq \frac{n r^n}{(r-1)^2} \|\Delta_r^2 y.\|_\infty,$$

$$\|\Delta_r y_n\|_X \leq \frac{r^n}{r-1} \|\Delta_r^2 y.\|_\infty.$$

---

[12]To prove Proposition 7.5.3, we will use the following properties:

- If $x = (x_n)$ is a sequence such that $\Delta_r x_n = y_n$, $x_0 = 0$ then

$$x_n = \sum_{i=0}^{n-1} r^i y_{n-1-i}.$$

- If $x = (x_n)$ is a sequence such that $\Delta_r^2 x_n = y_n$, $x_0 = x_1 = 0$, then

$$x_n = \sum_{i=0}^{n-2} \sum_{j=0}^{n-i-2} r^{i+j} y_{n-2-i-j} :$$

(c) The case $r = 1$. We get

$$\|y_n\|_X \leq \frac{n(n-1)}{2}\|\Delta_r^2 y.\|_\infty,$$

$$\|\Delta_r y_n\|_X \leq n\|\Delta_r^2 y.\|_\infty.$$

This completes the proof. $\qquad\qquad\qquad\qquad\qquad\qquad\qquad\qquad\qquad$ □

Denote by $\mathcal{B}^{\omega,r}$ the Banach space of all X-valued sequences $y = (y_n)_{n\in\mathbb{Z}_+}$ such that

$$\sup_{n\in\mathbb{Z}_+} \left(\frac{\|y_n\|_X}{c(r,n)} + \frac{\|\Delta_r y_n\|_X}{c(r,n+1)}\right) < +\infty,$$

endowed with the norm

$$\|y\|_\omega = \sup_{n\in\mathbb{Z}_+} \left(\frac{\|y_n\|_X}{c(r,n)} + \frac{\|\Delta_r y_n\|_X}{c(r,n+1)}\right). \qquad (7.5.5)$$

We note that the space X_ω of all X-valued sequences $y = (y_n)_{n\in\mathbb{Z}_+}$ such that $\sup_{n\in\mathbb{Z}_+} \frac{\|y_n\|_X}{c(r,n)} < +\infty$ is a Banach space endowed with the norm

$$\|y\|^\omega = \sup_{n\in\mathbb{Z}_+} \frac{\|y_n\|_X}{c(r,n)}.$$

From the following inequality[13]

$$\|y\|^\omega \leq \|y\|_\omega \leq (2+r)\|y\|^\omega$$

we deduce that $\mathcal{B}^{\omega,r}$ is a Banach space.

---

[13] We observe that

$$\|\Delta_r y_n\|_X \leq c(r,n+1)\frac{\|y_{n+1}\|_X}{c(r,n+1)} + rc(r,n)\frac{\|y_n\|_X}{c(r,n)}$$

$$\leq c(r,n+1)\sup_{n\in\mathbb{Z}_+}\frac{\|y_n\|_X}{c(r,n)} + rc(r,n)\sup_{n\in\mathbb{Z}_+}\frac{\|y_n\|_X}{c(r,n)}$$

$$\leq (1+r)c(r,n+1)\sup_{n\in\mathbb{Z}_+}\frac{\|y_n\|_X}{c(r,n)},$$

hence

$$\frac{\|y_n\|_X}{c(r,n)} + \frac{\|\Delta_r y_n\|_X}{c(r,n+1)} \leq (2+r)\sup_{n\in\mathbb{Z}_+}\frac{\|y_n\|_X}{c(r,n)}:$$

Proof of Theorem 7.5.1. Let us choose $m \in \mathbb{Z}_+$ such that

$$\vartheta := 2M \sum_{n=m}^{\infty} \rho_n c(r, n+1) < 1.$$

We denote by $\mathcal{B}_m^{\omega,r}$ the Banach space of all sequences $y = (y_n)_{n \in \mathbb{Z}_+}$ belonging to $\mathcal{B}^{\omega,r}$ such that $y_n = 0$ if $0 \leq n \leq m$ equipped with the norm $\| \cdot \|_\omega$.[14] We consider the operator Υ on $\mathcal{B}_m^{\omega,r}$ defined by

$$[\Upsilon y]_n = \begin{cases} 0, & \text{if } 0 \leq n \leq m, \\[2mm] \sum_{k=0}^{n-1-m} r^{k-1} S(k) G(n-1-k, \, y_{n-1-k}, \, \Delta_r y_{n-1-k}), & \text{if } n \geq m+1. \end{cases}$$

$$(7.5.6)$$

Next we show that Υ is well defined. Initially we observe that

$$\|(y_j, \Delta_r y_j)\|_{X \times X} = \|y_j\|_X + \|\Delta_r y_j\|_X$$

$$= c(r, j) \frac{\|y_j\|_X}{c(r, j)} + c(r, j+1) \frac{\|\Delta_r y_j\|_X}{c(r, j+1)}$$

$$\leq c(r, j+1) \left(\frac{\|y_j\|_X}{c(r, j)} + \frac{\|\Delta_r y_j\|_X}{c(r, j+1)} \right)$$

$$\leq c(r, j+1) \sup_{j \in \mathbb{Z}_+} \left(\frac{\|y_j\|_X}{c(r, j)} + \frac{\|\Delta_r y_j\|_X}{c(r, j+1)} \right)$$

$$= c(r, j+1) \|\|y\|\|_\omega.$$

$$(7.5.7)$$

Keeping in mind conditions (a)** and (b)**, we have the following estimates:

$$\|[\Upsilon y]_n\|_X \leq \sum_{k=0}^{n-1-m} \|r^{k-1} S(k)\| \|G(n-1-k, \, y_{n-1-k}, \, \Delta_r y_{n-1-k})\|_X$$

$$\leq M \sum_{k=0}^{n-1-m} c(r, k) \|G(n-1-k, \, y_{n-1-k}, \, \Delta_r y_{n-1-k})\|_X$$

---

[14] We observe that $\mathcal{B}_m^{\omega,r}$ is a closed subspace of $\mathcal{B}^{\omega,r}$.

$$= M \sum_{j=m}^{n-1} c(r, n-1-j) \| G(j, y_j, \Delta_r y_j) \|_X$$

$$\leq Mc(r, n) \sum_{j=m}^{n-1} \| G(j, y_j, \Delta_r y_j) - G(j, 0, 0) \|_X$$

$$+ Mc(r, n) \sum_{j=m}^{n-1} \| G(j, 0, 0) \|_X.$$

Using (7.5.7) we get

$$\frac{\| [\Upsilon y]_n \|_X}{c(r, n)} \leq M \sum_{j=m}^{n-1} \rho_j \| (y_j, \Delta_r y_j) \|_{X \times X} + M \| G(\cdot, 0, 0) \|_1$$

$$\leq M \sum_{j=m}^{\infty} \rho_j c(r, j+1) \| y \|_\omega + M \| G(\cdot, 0, 0) \|_1. \qquad (7.5.8)$$

On the other hand, we observe that

$$\Delta_r [\Upsilon y]_n = \sum_{k=1}^{n-m} r^{k-1} S(k) G(n-k, y_{n-k}, \Delta_r y_{n-k})$$

$$- \sum_{k=0}^{n-1-m} r^k S(k) G(n-1-k, y_{n-1-k}, \Delta_r y_{n-1-k})$$

$$= \sum_{j=0}^{n-1-m} r^j (S(j+1) - S(j)) G(n-1-j, y_{n-1-j}, \Delta_r y_{n-1-j})$$

$$= \sum_{j=0}^{n-1-m} r^j C(j) G(n-1-j, y_{n-1-j}, \Delta_r y_{n-1-j}). \qquad (7.5.9)$$

Therefore

$$\frac{\| \Delta_r [\Upsilon y]_n \|_X}{c(r, n+1)} \leq M \sum_{j=m}^{\infty} \rho_j c(r, j+1) \| y \|_\omega + M \| G(\cdot, 0, 0) \|_1. \qquad (7.5.10)$$

From estimates (7.5.8) and (7.5.10), we obtain $\Upsilon y \in \mathcal{B}_m^{\omega,r}$.[15]

We find that the operator Υ is a ϑ-contraction. In fact for y^1 and y^2 in $\mathcal{B}_m^{\omega,r}$ we have the following estimates:

$$\|[\Upsilon y^1]_n - [\Upsilon y^2]_n\|_X \leq \sum_{k=0}^{n-1-m} \|r^{k-1} \mathcal{S}(k)\| \|G(n-1-k, y_{n-1-k}^1, \Delta_r y_{n-1-k}^1)$$

$$-G(n-1-k, y_{n-1-k}^2, \Delta_r y_{n-1-k}^2)\|_X$$

$$\leq M \sum_{k=0}^{n-1-m} c(r,k) \|G(n-1-k, y_{n-1-k}^1, \Delta_r y_{n-1-k}^1)$$

$$-G(n-1-k, y_{n-1-k}^2, \Delta_r y_{n-1-k}^2)\|_X$$

$$= M \sum_{j=m}^{n-1} c(r, n-1-j) \|G(j, y_j^1, \Delta_r y_j^1)$$

$$-G(j, y_j^2, \Delta_r y_j^2)\|_X$$

$$\leq Mc(r,n) \sum_{j=m}^{n-1} \rho_j \|((y^1 - y^2)_j, \Delta_r(y^1 - y^2)_j)\|_{X \times X}$$

$$\leq Mc(r,n) \sum_{j=m}^{n-1} \rho_j c(r, j+1) \|y^1 - y^2\|_\omega.$$

Hence

$$\frac{\|[\Upsilon y^1]_n - [\Upsilon y^2]_n\|_X}{c(r,n)} \leq M \sum_{j=m}^{n-1} \rho_j c(r, j+1) \|y^1 - y^2\|_\omega$$

$$\leq \left(M \sum_{j=m}^{\infty} \rho_j c(r, j+1) \right) \|y^1 - y^2\|_\omega.$$

$$(7.5.11)$$

---

[15]Note that $\|\Upsilon y\|_\omega \leq \|y\|_\omega + 2M \|G(\cdot, 0, 0)\|_1$.

From (7.5.9) we get

$$\Delta_r[\Upsilon y^1]_n - \Delta_r[\Upsilon y^2]_n = \sum_{j=0}^{n-1-m} r^j C(j)(G(n-1-j, y^1_{n-1-j}, \Delta_r y^1_{n-1-j})$$

$$-G(n-1-j, y^2_{n-1-j}, \Delta_r y^2_{n-1-j})).$$

Hence

$$\frac{\|\Delta[\Upsilon y^1]_n - \Delta[\Upsilon y^2]_n\|_X}{c(r,n+1)} \leq \Big(M \sum_{j=m}^{\infty} \rho_j c(r,j+1)\Big)\|y^1 - y^2\|_\omega.$$

(7.5.12)

From (7.5.11) and (7.5.12) we obtain

$$\|\Upsilon y^1 - \Upsilon y^2\|_\omega \leq 2M \sum_{j=m}^{\infty} \rho_j c(r,j+1)\|y^1 - y^2\|_\omega.$$

Thus our assertion follows.

Let y be the unique fixed point of Υ; from (7.5.7), (7.5.8), and (7.5.10), we have the following estimates:

$$\frac{\|y_n\|_X}{c(r,n)} \leq M \sum_{j=m}^{n-1} \rho_j \|(y_j, \Delta_r y_j)\|_{X\times X} + M\|G(\cdot,0,0)\|_1$$

$$\leq M \sum_{j=m}^{n-1} \rho_j c(r,j+1)\Big(\frac{\|y_j\|_X}{c(r,j)} + \frac{\|\Delta_r y_j\|_X}{c(r,j+1)}\Big)$$

$$+M\|G(\cdot,0,0)\|_1 \tag{7.5.13}$$

and

$$\frac{\|\Delta_r y_n\|_X}{c(r,n)} \leq M \sum_{j=m}^{n-1} \rho_j c(r,j+1)\Big(\frac{\|y_j\|_X}{c(r,j)} + \frac{\|\Delta_r y_j\|_X}{c(r,j+1)}\Big)$$

$$+M\|G(\cdot,0,0)\|_1. \tag{7.5.14}$$

Now we introduce the function

$$u_n := \frac{||y_n||_X}{c(r,n)} + \frac{||\Delta_r y_n||_X}{c(r,n+1)}.$$

In view of (7.5.13) and (7.5.14) we obtain

$$u_n \leq 2M ||G(\cdot,0,0)||_1 + 2M \sum_{j=m}^{n-1} \rho_j c(r, j+1) u_j.$$

The discrete Gronwall's inequality applied to the above inequality yields

$$u_n \leq M ||G(\cdot,0,0)||_1 \prod_{j=m}^{n-1} \left(1 + 2M\rho_j c(r, j+1)\right)$$

$$\leq M ||G(\cdot,0,0)||_1 e^{2M \sum_{j=m}^{n-1} \rho_j c(r,j+1)}$$

$$\leq 2M ||G(\cdot,0,0)||_1 e^{2M ||c(r,\cdot+1)\rho\cdot||_1},$$

which completes the proof of Theorem 7.5.1. □

Example 7.5.4. Let f be defined and satisfy a Lipschitz condition in $\mathbb{R}$ and let $q = (q_n)$ be a complex sequence such that $\sum_{n=0}^{\infty} |q_n| 2^{2n} < +\infty$. Then for the evolution problem:

$$x_{n+2} - 4x_{n+1} + 4x_n = q_n f(x_n), \ n \in \mathbb{Z}_+. \tag{7.5.15}$$

there is a unique weighted bounded solution $x = (x_n)$ such that for $n \geq m \in \mathbb{Z}_+$

$$\sup_{n \in \mathbb{Z}_+} \frac{1}{4^{n+1}} \left(|x_n| + |x_{n+1} - 2x_n|\right) < +\infty. \tag{7.5.16}$$

In fact, we consider $X = \mathbb{R}$, $T = I$, $r = 2$, $c(2, n) = 2^{2n+1}$, $G(n, z_1, z_2) = q_n f(z_1)$, so that condition $(A_1)^{**}$ is fulfilled and that $M = \frac{1}{2}$ in (b)**. Then by Theorem 7.5.1 there is a positive constant $m \in \mathbb{Z}_+$ and a unique weighted bounded solution $x = (x_n)$ of (7.5.15) for $n \geq m$ such that (7.5.16) holds.

To conclude this section, we establish a version of Theorem 7.5.1 which enable us to consider locally Lipschitz perturbations of (6.3.1). In the process of obtaining this result we require the assumption:

Assumption $(A_2)^{**}$: Let $(c(r, n))_n$ be a positive sequence which is nondecreasing with respect to the second variable. Suppose that the following conditions hold:

(i)$^\natural$ The function $G : \mathbb{Z}_+ \times X \times X \longrightarrow X$ satisfies a locally Lipschitz condition in $X \times X$, i.e., for each positive number R, for all $n \in \mathbb{Z}_+$, and $z, w \in X \times X$, $||z||_{X \times X} \le R$, $||w||_{X \times X} \le R$,

$$||G(n, z) - G(n, w)||_X \le l(n, R)||z - w||_{X \times X},$$

where $l : \mathbb{Z}_+ \times [0, \infty) \longrightarrow [0, \infty)$ is a nondecreasing function with respect to the second variable.

(ii)$^\natural$ There is a positive number v such that

$$\sum_{n=0}^{\infty} l(n, c(r, n + 1)v)c(r, n + 1) < \infty.$$

(iii)$^\natural$ $G(\cdot, 0, 0) \in l_1(\mathbb{Z}_+; X)$.

We have the following local version of Theorem 7.5.1.

Theorem 7.5.5. *Let X be a Banach space and $T \in \mathcal{B}(X)$. Assume that $(A_2)^{**}$ and the condition $(b)^{**}$ of Theorem 7.5.1 are fulfilled. Then, there is a positive constant $m \in \mathbb{Z}_+$ and a weighted bounded solution $y = (y_n)$ of (6.3.1) for $n \ge m$ such that $y_n = 0$, $0 \le n \le m$ and $||y||_\omega \le v$, where v is the constant of condition $(ii)^\natural$ in $(A_2)^{**}$.*

Proof. Let $\beta \in (0, 1)$. Using $(iii)^\natural$ and $(ii)^\natural$ respectively, there is $m \in \mathbb{Z}_+$ such that

$$2M \sum_{n=m}^{\infty} ||G(n, 0, 0)||_X \le \beta v, \qquad (7.5.17)$$

and

$$\mathcal{T} := \beta + 2M \sum_{n=m}^{\infty} l(n, c(r, n + 1)v)c(r, n + 1) < 1, \qquad (7.5.18)$$

where M is given in Theorem 7.5.1.

We denote by $\mathcal{B}_m^{\omega,r}[v]$ the ball $||y||_\omega \le v$ in $\mathcal{B}_m^{\omega,r}$. Let y be a sequence in $\mathcal{B}_m^{\omega,r}[v]$. Using (7.5.7), we have

$$||[\Upsilon y]_n||_X \le Mc(r, n)\left(\sum_{j=m}^{n-1} l(j, c(r, j + 1)v)c(r, j + 1)||y||_\omega \right.$$

$$\left. + \sum_{j=m}^{n-1} ||G(j, 0, 0)||_X \right),$$

$$\|\Delta_r[\Upsilon y]_n\|_X \leq Mc(r, n+1)\Big(\sum_{j=m}^{n-1} l(j, c(r, j+1)v)c(r, j+1)\|y\|_\omega$$

$$+ \sum_{j=m}^{n-1} \|G(j, 0, 0)\|_X \Big),$$

whence

$$\frac{\|[\Upsilon y]_n\|_X}{c(r, n)} \leq M \sum_{j=m}^{\infty} l(j, c(r, j+1)v)c(r, j+1)v + M \sum_{j=m}^{\infty} \|G(j, 0, 0)\|_X,$$

$$(7.5.19)$$

$$\frac{\|\Delta[\Upsilon y]_n\|_X}{c(r, n+1)} \leq M \sum_{j=m}^{\infty} l(j, c(r, j+1)v)c(r, j+1)v + M \sum_{j=m}^{\infty} \|G(j, 0, 0)\|_X.$$

$$(7.5.20)$$

Replacing (7.5.17) and (7.5.18) in (7.5.19) and (7.5.20) respectively,

$$\|\Upsilon y\|_\omega \leq 2M \sum_{j=m}^{\infty} l(j, c(r, j+1)v)c(r, j+1)v + 2M \sum_{j=m}^{\infty} \|G(j, 0, 0)\|_X.$$

$$\leq (\mathcal{T} - \beta)v + \beta v = \mathcal{T}v < v.$$

On the other hand, we obtain

$$\|\Upsilon y^1 - \Upsilon y^2\|_\omega \leq 2M \sum_{j=m}^{\infty} l(j, c(r, j+1)v)c(r, j+1)\|y^1 - y^2\|_\omega$$

$$\leq (\mathcal{T} - \beta)\|y^1 - y^2\|_\omega.$$

Since $(\mathcal{T} - \beta) < 1$, Υ is a contraction. $\qquad\square$

7.6 Asymptotic Behavior

In many problems one encounters quantities that arise naturally and are worth considering, but for which no exact formula is known. In such cases we can derive approximations for these quantities when the argument tends to infinity. In applications such approximations are often just as useful as an exact formula would be. Asymptotic analysis is a method of describing limiting behavior and a key tool for exploring the evolution equations which arise in the mathematical modeling of real-world phenomena.

In this section, we give the asymptotic behavior of the solutions of (6.3.1).

Theorem 7.6.1. *Let X be a Banach space and $T \in \mathcal{B}(X)$. Assume that the following conditions hold:*

(C_1) Assumption $(A_1)^{**}$ is fulfilled (see Sect. 7.5).

(C_2) $\displaystyle\lim_{n\to\infty} \frac{r^{n-1}S(n)}{c(r,n-1)} = 0$ *in $\mathcal{B}(X)$, where $c(r,n)$ is given by $(A_1)^{**}$.*

Then, there is a positive constant $m \in \mathbb{Z}_+$ and a unique solution $y = (y_n)$ of (6.3.1) for $n \geq m$ such that $y_1 = \cdots = y_m = 0$ and $y_n = o(c(r,n))$ as $n \to \infty$.

Remark 7.6.2. In the following cases (C_2) is fulfilled:

(i) $T = I$ and $\displaystyle\lim_{n\to\infty} \frac{nr^{n-1}}{c(r,n-1)} = 0$.

(ii) If (5.5.8) has a discrete maximal l_∞-regularity [see Remark 7.5.2(ii)] with $T \in \mathcal{B}(X)$ such that $1 \in \rho(T)$ and $c(r,n) := \gamma(r,n)$, $r > 1$ [see (5.5.16)], then (C_2) is true for all $r > 1$. In fact, since (5.5.8) has a discrete maximal l_∞-regularity, we get

$$\sup_{n\in\mathbb{Z}_+} \left\| \sum_{k=0}^n (I-T)r^{k+1}S(k)f_{n-k} \right\|_X \leq \|\mathcal{K}_\infty^r\|_{\mathcal{B}(L_\infty(\mathbb{Z}_+,X))}\|f\|_\infty.$$

Let $x \in X$ be an arbitrary vector and we define

$$f(n) = \begin{cases} x, & n = 0, \\ 0, & n \neq 0. \end{cases}$$

Then

$$\sup_{n\in\mathbb{Z}_+} \|(I-T)r^{n+1}S(n)x\|_X \leq \|\mathcal{K}_\infty^r\|_{\mathcal{B}(L_\infty(\mathbb{Z}_+,X))}\|x\|_X.$$

Therefore,

$$\|(I-T)r^{n+1}S(n)\| \leq \|\mathcal{K}_\infty^r\|_{\mathcal{B}(L_\infty(\mathbb{Z}_+,X))}, \text{ for all } n \in \mathbb{Z}_+.$$

We note that

$$\|r^{n+1}S(n)\| = \|(I-T)^{-1}(I-T)r^{n+1}S(n)\|$$
$$= \|(I-T)^{-1}\|\|(I-T)r^{n+1}S(n)\|$$
$$= \|(I-T)^{-1}\|\|\mathcal{K}_\infty^r\|_{\mathcal{B}(L_\infty(\mathbb{Z}_+,X))},$$

hence

$$||r^{\bullet+1}S(\bullet)||_\infty := \sup_{n\in\mathbb{Z}_+} ||r^{n+1}S(n)|| < \infty.$$

For $r > 1$ we get

$$\left\|\frac{r^{n-1}S(n)}{\gamma(r, n-1)}\right\| \le \frac{r^{-2}}{\gamma(r, n-1)} \sup_{n\in\mathbb{Z}_+} r^{n+1}||S(n)||$$

$$\le \frac{r^{-2}}{(n-1)r^{n-1}}||r^{\bullet+1}S(\bullet)||_\infty$$

Therefore

$$\lim_{n\to\infty} \frac{r^{n-1}S(n)}{\gamma(r, n-1)} = 0.$$

We denote by $\mathcal{B}_\infty^{\omega,r}$ the Banach space of all weighted convergent functions $\xi \in \mathcal{B}_m^{\omega,r}$ that is for which the limit

$$Z_\infty^{\omega,r}(\xi) := \lim_{n\to\infty} \frac{\xi(n)}{c(r, n)}$$

exists, endowed with the norm $|| \cdot ||_\omega$ [see (7.5.5)].

Theorem 7.6.1 is a consequence of Proposition 1.4.4 and Banach's contraction principle applied to the operator $\tilde{\Upsilon}$ on $\mathcal{B}_\infty^{\omega,r}$ defined by (7.5.6).

Proof of Theorem 7.6.1. Let us choose $m \in \mathbb{Z}_+$ such that

$$\left(\left\|\frac{r^{\bullet-1}S(\bullet)}{c(r, \bullet - 1)}\right\|_\infty + \left\|\frac{r^{\bullet}C(\bullet)}{c(r, \bullet)}\right\|_\infty\right)\left(\sum_{j=m}^\infty \rho_j c(r, j+1)\right) < 1. \tag{7.6.1}$$

We consider the operator $\tilde{\Upsilon}$ on $\mathcal{B}_\infty^{\omega,r}$ defined by (7.5.6). We show that $\tilde{\Upsilon}$ is well defined. We argue as follows: let n_1 be a natural number large enough and let y be in $\mathcal{B}_\infty^{\omega,r}$ arbitrary and $n \ge n_1$. Keeping in mind Proposition 1.4.4 one can deduce the following identity:

$$[\tilde{\Upsilon} y]_n = \sum_{k=1}^{n-1-m} r^{k-1} \mathcal{S}(k) G(n-1-k,\ y_{n-1-k},\ \Delta_r y_{n-1-k})$$

$$= \sum_{j=m}^{n-2} r^{n-2-j} \mathcal{S}(n-1-j) G(j,\ y_j,\ \Delta_r y_j)$$

$$= \sum_{j=m}^{n_1-1} r^{n-2-j} \mathcal{S}(n-1-j) G(j,\ y_j, \Delta_r y_j)$$

$$+ \sum_{j=n_1}^{n-2} r^{n-2-j} \mathcal{S}(n-1-j) G(j,\ y_j, \Delta_r y_j)$$

$$= \sum_{j=m}^{n_1-1} r^{n-2-j} \Big(\mathcal{S}(n-1-n_1) \mathcal{C}(n_1-j)$$

$$+ \mathcal{S}(n_1-j) \mathcal{C}(n-1-n_1) \Big) G(j,\ y_j, \Delta_r y_j)$$

$$+ \sum_{j=n_1}^{n-2} r^{n-2-j} \mathcal{S}(n-1-j) G(j,\ y_j, \Delta_r y_j)$$

$$= I_1^n + I_2^n + I_3^n,$$

where

$$I_1^n := r^{n-2-n_1} \mathcal{S}(n-1-n_1) \sum_{j=m}^{n_1-1} r^{n_1-j} \mathcal{C}(n_1-j) G(j,\ y_j, \Delta_r y_j),$$

$$I_2^n := r^{n-1-n_1} \mathcal{C}(n-1-n_1) \sum_{j=m}^{n_1-1} r^{n_1-j-1} \mathcal{S}(n_1-j) G(j,\ y_j, \Delta_r y_j),$$

$$I_3^n := \sum_{j=n_1}^{n-2} r^{n-2-j} \mathcal{S}(n-1-j) G(j,\ y_j, \Delta_r y_j).$$

Taking into account (7.5.7) we have the following estimates:

$$\|I_1^n\|_X \leq r^{n-2-n_1} \|\mathcal{S}(n-1-n_1)\| \sum_{j=m}^{n_1-1} r^{n_1-j} \|\mathcal{C}(n_1-j)\|$$

$$\times \|G(j, y_j, \Delta_r y_j)\|_X$$

$$\leq r^{n-2-n_1} \|\mathcal{S}(n-1-n_1)\| \sum_{j=m}^{n_1-1} r^{n_1-j} \|\mathcal{C}(n_1-j)\|$$

$$\times \Big(\|G(j, y_j, \Delta_r y_j) - G(j,0,0)\|_X + \|G(j,0,0)\|_X \Big)$$

$$\leq r^{n-2-n_1} \|\mathcal{S}(n-1-n_1)\| \sum_{j=m}^{n_1-1} r^{n_1-j} \frac{\|\mathcal{C}(n_1-j)\|}{c(r, n_1-j)} c(r, n_1-j)$$

$$\times \Big(\rho_j \|(y_j, \Delta_r y_j)\|_{X \times X} + \|G(j,0,0)\|_X \Big)$$

$$\leq r^{n-2-n_1} \|\mathcal{S}(n-1-n_1)\| \left\| \frac{r^\bullet \mathcal{C}(\bullet)}{c(r,\bullet)} \right\|_\infty c(r, n_1)$$

$$\times \Big(\Big(\sum_{j=m}^{n_1-1} \rho_j c(r, j+1) \Big) \|\|y\|\|_\omega + \Big(\sum_{j=m}^{n_1-1} \|G(j,0,0)\|_X \Big) \Big)$$

$$\leq r^{n-2-n_1} \|\mathcal{S}(n-1-n_1)\| \left\| \frac{r^\bullet \mathcal{C}(\bullet)}{c(r,\bullet)} \right\|_\infty c(r, n_1)$$

$$\times \Big(\|\rho_\bullet c(r, \bullet+1)\|_1 \|\|y\|\|_\omega + \|G(\bullet,0,0)\|_1 \Big),$$

$$(7.6.2)$$

$$\|I_2^n\|_X \leq r^{n-1-n_1} \|\mathcal{C}(n-1-n_1)\| \sum_{j=m}^{n_1-1} r^{n_1-j-1} \|\mathcal{S}(n_1-j)\|$$

$$\times \Big(\rho_j c(r, j+1) \|\|y\|\|_\omega + \|G(j,0,0)\|_X \Big)$$

$$\leq r^{n-1-n_1} \left\| \mathcal{C}(n-1-n_1) \right\| \left\| \frac{r^{\bullet-1} \mathcal{S}(\bullet)}{c(r, \bullet-1)} \right\|_\infty c(r, n_1)$$

$$\times \left(\left\| \rho_\bullet c(r, \bullet+1) \right\|_1 |||y|||_\omega + ||G(\bullet, 0, 0)||_1 \right),$$

$$(7.6.3)$$

$$||I_3^n||_X \leq \sum_{j=n_1}^{n-2} r^{n-2-j} \left\| \mathcal{S}(n-1-j) \right\| \left(\rho_j c(r, j+1) |||y|||_\omega \right.$$

$$\left. + \left\| G(j, 0, 0) \right\|_X \right)$$

$$\leq \left\| \frac{r^{\bullet-1} \mathcal{S}(\bullet)}{c(r, \bullet-1)} \right\|_\infty \sum_{j=n_1}^{n-2} c(r, n-2-j) \left(\rho_j c(r, j+1) |||y|||_\omega \right.$$

$$\left. + \left\| G(j, 0, 0) \right\|_X \right)$$

$$\leq \left\| \frac{r^{\bullet-1} \mathcal{S}(\bullet)}{c(r, \bullet-1)} \right\|_\infty c(r, n) \left(\left(\sum_{j=n_1}^{\infty} \rho_j c(r, j+1) \right) |||y|||_\omega \right.$$

$$\left. + \sum_{j=n_1}^{\infty} \left\| G(j, 0, 0) \right\|_X \right).$$

$$(7.6.4)$$

We infer from (7.6.2) and (7.6.3) that there exists a positive constant $\tilde{C}$ independent of n so that

$$\frac{||I_1^n||_X}{c(r, n)} \leq \tilde{C} r^{n-1-n_1-1} \frac{||\mathcal{S}(n-1-n_1)||}{c(r, n-1-n_1)}, \qquad (7.6.5)$$

and

$$\frac{||I_2^n||_X}{c(r, n)} \leq \tilde{C} r^{n-1-n_1} \frac{||\mathcal{C}(n-1-n_1)||}{c(r, n-1-n_1)}. \qquad (7.6.6)$$

On the other hand, taking advantage of condition (C_2), we deduce that $\lim_{n\to\infty} \frac{r^n C(n)}{c(r,n)} = 0$ in $\mathcal{B}(X)$. Indeed from Proposition 1.4.4 we have

$$\lim_{n\to\infty} \frac{r^n C(n)}{c(r,n)} = \lim_{n\to\infty} \frac{r^n \Delta S(n)}{c(r,n)}$$

$$= \lim_{n\to\infty} \frac{r^n S(n+1)}{c(r,n)} - \lim_{n\to\infty} \frac{r^n S(n)}{c(r,n)}$$

$$= \lim_{n\to\infty} \frac{r^{n-1}S(n)}{c(r,n-1)} - r\lim_{n\to\infty} \frac{r^{n-1}S(n)}{c(r,n)} = 0. \qquad (7.6.7)$$

We infer from (7.6.5)–(7.6.7) and (7.6.4) that

$$I_i^n = o(c(r,n)) \text{ as } n \to \infty, \ i = 1,2,3.$$

Hence

$$Z_\infty^{\omega,r}(\tilde\Upsilon y) = 0.$$

This allows us to prove that the space $\mathcal{B}_\infty^{\omega,r}$ is invariant under $\tilde\Upsilon$.

Let y^1 and y^2 be in $\mathcal{B}_\infty^{\omega,r}$. In view of Assumption $(A_1)^{**}$ and (7.5.7), we have initially for $n \geq m+1$

$$\|[\tilde\Upsilon y^1]_n - [\tilde\Upsilon y^2]_n\|_X$$

$$\leq \sum_{k=0}^{n-1-m} r^{k-1}\|S(k)\|\|G(n-1-k,\ y_{n-1-k}^1,\ \Delta_r y_{n-1-k}^1)$$

$$-G(n-1-k,\ y_{n-1-k}^2,\ \Delta_r y_{n-1-k}^2)\|_X$$

$$\leq \sum_{k=0}^{n-1-m} r^{k-1}\|S(k)\|\rho_{n-1-k}\|((y^1-y^2)_{n-1-k},\ \Delta_r(y^1-y^2)_{n-1-k})\|_{X\times X}$$

$$\leq \sum_{k=0}^{n-1-m} r^{k-1}\|S(k)\|\rho_{n-1-k}c(r,n-k)\|\|y^1-y^2\|\|_\omega$$

$$\leq \left\|\frac{r^{\bullet-1}S(\bullet)}{c(r,\bullet-1)}\right\|_\infty c(r,n)\left(\sum_{j=m}^{n-1}\rho_j c(r,j+1)\right)\|\|y^1-y^2\|\|_\omega.$$

Hence, for all $n \in \mathbb{Z}_+$, we obtain

$$\frac{\|[\tilde\Upsilon y^1]_n - [\tilde\Upsilon y^2]_n\|_X}{c(r,n)} \leq \left\|\frac{r^{\bullet-1}S(\bullet)}{c(r,\bullet-1)}\right\|_\infty\left(\sum_{j=m}^{\infty}\rho_j c(r,j+1)\right)\|\|y^1-y^2\|\|_\omega.$$

$$(7.6.8)$$

To prove that $\tilde{\Upsilon}$ is a contraction, we make use of the following identity:

$$
\Delta_r[\tilde{\Upsilon} y]_n = \begin{cases} G(m, 0, y_{m+1}), & \text{if } n = m, \\[2mm] \displaystyle\sum_{j=0}^{n-1-m} r^j C(j) G(n-1-j, \ y_{n-1-j}, \ \Delta_r y_{n-1-j}), & \text{if } n \geq m+1. \end{cases}
$$

(7.6.9)

Indeed we have for $n \geq m+1$

$$
\Delta_r[\tilde{\Upsilon} y]_n = [\tilde{\Upsilon} y]_{n+1} - r[\tilde{\Upsilon} y]_n
$$

$$
= \sum_{j=0}^{n-1-m} r^j S(j+1) G(n-1-j, \ y_{n-1-j}, \ \Delta_r y_{n-1-j})
$$

$$
- \sum_{j=0}^{n-1-m} r^j S(j) G(n-1-j, \ y_{n-1-j}, \ \Delta_r y_{n-1-j})
$$

$$
= \sum_{j=0}^{n-1-m} r^j (S(j+1) - S(j)) G(n-1-j, \ y_{n-1-j}, \ \Delta_r y_{n-1-j})
$$

$$
= \sum_{j=0}^{n-1-m} r^j (\Delta S(j)) G(n-1-j, \ y_{n-1-j}, \ \Delta_r y_{n-1-j})
$$

$$
= \sum_{j=0}^{n-1-m} r^j C(j) G(n-1-j, \ y_{n-1-j}, \ \Delta_r y_{n-1-j}).
$$

Hence we find that the operator $\tilde{\Upsilon}$ satisfies (7.6.9).

Next, we note that

$$
\begin{aligned}
\|\Delta_r[\tilde{\Upsilon} y^1]_m - \Delta_r[\tilde{\Upsilon} y^2]_m\|_X &= \|G(m, 0, y^1_{m+1}) - G(m, 0, y^2_{m+1})\|_X \\[1mm]
&\leq \rho_m \|(0, y^1_{m+1} - y^2_{m+1})\|_{X \times X} \\[1mm]
&= \rho_m c(r, m+1) \frac{\|y^1_{m+1} - y^2_{m+1}\|_X}{c(r, m+1)} \\[1mm]
&\leq \rho_m c(r, m+1) \||y^1 - y^2\||_\omega \\[1mm]
&= \frac{1}{c(r, 0)} \rho_m c(r, 0) c(r, m+1) \||y^1 - y^2\||_\omega \\[1mm]
&\leq \left\| \frac{r^\bullet C(\bullet)}{c(r, \bullet)} \right\|_\infty \rho_m c(r, m+1) c(r, m+1) \||y^1 - y^2\||_\omega.
\end{aligned}
$$

Hence

$$\frac{\|\Delta_r[\tilde{\Upsilon}y^1]_m - \Delta_r[\tilde{\Upsilon}y^2]_m\|_X}{c(r, m+1)} \le \left\|\frac{r^\bullet C(\bullet)}{c(r, \bullet)}\right\|_\infty \left(\sum_{j=m}^\infty \rho_j c(r, j+1)\right) \|\|y^1 - y^2\|\|_\omega.$$

$$(7.6.10)$$

Using the identity (7.6.9), we have the following estimates for $n \ge m + 1$:

$$\|\Delta_r[\tilde{\Upsilon}y^1]_n - \Delta_r[\tilde{\Upsilon}y^2]_n\|_X$$

$$\le \sum_{k=m}^{n-1} r^{n-1-k}\|C(n-1-k)\|\|\rho_k\|\|((y^1 - y^2)_k, \Delta_r(y^1 - y^2)_k)\|_{X \times X}$$

$$\le \left\|\frac{r^\bullet C(\bullet)}{c(r, \bullet)}\right\|_\infty c(r, n+1)\left(\sum_{j=m}^{n-1} \rho_j c(r, j+1)\right) \|\|y^1 - y^2\|\|_\omega,$$

whence

$$\frac{\|\Delta_r[\tilde{\Upsilon}y^1]_n - \Delta_r[\tilde{\Upsilon}y^2]_n\|_X}{c(r, n+1)} \le \left\|\frac{r^\bullet C(\bullet)}{c(r, \bullet)}\right\|_\infty \left(\sum_{j=m}^\infty \rho_j c(r, j+1)\right) \|\|y^1 - y^2\|\|_\omega.$$

$$(7.6.11)$$

Using the estimates (7.6.8), (7.6.10), and (7.6.11), we obtain

$$\|\|\tilde{\Upsilon}y^1 - \tilde{\Upsilon}y^2\|\|_\omega \le \left(\left\|\frac{r^{\bullet-1}S(\bullet)}{c(r, \bullet-1)}\right\|_\infty + \left\|\frac{r^\bullet C(\bullet)}{c(r, \bullet)}\right\|_\infty\right)\left(\sum_{j=m}^\infty \rho_j c(r, j+1)\right)\|\|y^1 - y^2\|\|_\omega.$$

By (7.6.1), $\tilde{\Upsilon}$ has a unique fixed point $y \in \mathcal{B}_\infty^{\omega,r}$ such that $Z_\infty^{\omega,r}(\tilde{\Upsilon}y) = 0$. This completes the proof of the theorem. □

7.7 Boundedness for RFDE via Maximal Regularity

The methods presented in Chap. 4 can be used to study the existence and uniqueness of bounded solutions which are in l_p for semilinear functional difference equations with infinite delay (*RFDE* for short).

Several aspects of the theory of functional difference equations can be understood as a proper generalization of the theory of ordinary difference equations. However, the fact that the state space for functional difference equations is infinite dimensional requires the development of methods and techniques coming from functional analysis (e.g., theory of semigroups of operators on Banach spaces, spectral theory, etc.)

Besides its theoretical interest, the study of abstract retarded functional difference equations in phase space has great importance in applications. Because of these reasons the theory of difference equations with infinite delay has drawn the attention of several authors. In fact, properties of the solutions have been studied in several contexts. For example, invariant manifolds theory [140], convergence theory [50, 56, 57], discrete maximal regularity [59], asymptotic behavior [51, 58, 76, 82, 141, 142], exponential dichotomy [36, 167], robustness [36], stability [83, 84, 150], and periodicity [2, 15, 40, 60, 63, 97, 136, 139, 151, 170–173, 178].

However, until now the literature concerning discrete maximal regularity for functional difference equations with infinite delay is too incipient and should be developed, so that it can produce progress in the theory of Volterra difference equations with infinite delay (see Sect. 7.9).

In the present section we are concerned with the study of the existence of bounded solutions for the semilinear problem

$$x(n + 1) = L(n, x_n) + f(n, x_\bullet), \quad n \geq 0, \tag{7.7.1}$$

by means of the knowledge of maximal regularity properties for the retarded linear functional equation

$$x(n + 1) = L(n, x_n), \quad n \geq 0, \tag{7.7.2}$$

where $L : \mathbb{N} \times \mathcal{P}_{ps} \longrightarrow \mathbb{C}^r$ is a bounded linear map with respect to the second variable, $x_\bullet$ denotes the $\mathcal{P}_{ps}$-valued function defined by $n \longrightarrow x_n$, here $x_n : \mathbb{Z}^- \longrightarrow \mathbb{C}^r$ is the history function which is defined by $x_n(s) = x(n + s)$ for all $s \in \mathbb{Z}^-$, and $\mathcal{P}_{ps}$ denotes an abstract phase space which is defined axiomatically.

Following the terminology used in Murakami [148] the phase space $\mathcal{P}_{ps} = \mathcal{P}_{ps}(\mathbb{Z}^-, \mathbb{C}^r)$ is a Banach space (with norm denoted by $||\cdot||_{\mathcal{P}_{ps}}$) which is a subfamily of functions from $\mathbb{Z}^-$ into $\mathbb{C}^r$ and it is assumed to satisfy the following axioms:

(PS_1) There is a positive constant $J > 0$ and nonnegative functions $N(\cdot)$ and $M(\cdot)$ on $\mathbb{Z}^+$ with the property that if $x : \mathbb{Z} \longrightarrow \mathbb{C}^r$ is a function such that $x_0 \in \mathcal{P}_{ps}$, then for all $n \in \mathbb{Z}^+$

    (i) $x_n \in \mathcal{P}_{ps}$.
    (ii) $J|x(n)| \leq ||x_n||_{\mathcal{P}_{ps}} \leq N(n) \sup_{0 \leq s \leq n} |x(s)| + M(n)||x_0||_{\mathcal{P}_{ps}}$.

(PS_2) The inclusion map $i : (B(\mathbb{Z}^-, \mathbb{C}^r), ||\cdot||_\infty) \longrightarrow (\mathcal{P}_{ps}, ||\cdot||_{\mathcal{P}_{ps}})$ is continuous, i.e., there is a constant $K_{\mathcal{P}_{ps}} \geq 0$ such that $||\varphi||_{\mathcal{P}_{ps}} \leq K_{\mathcal{P}_{ps}}||\varphi||_\infty$, for all $\varphi \in B(\mathbb{Z}^-, \mathbb{C}^r)$ (where $B(\mathbb{Z}^-, \mathbb{C}^r)$ represents the set of all bounded functions from $\mathbb{Z}^-$ into $\mathbb{C}^r$).

For any $n \geq \tau$ we define the operator $T(n, \tau): \mathcal{P}_{ps} \to \mathcal{P}_{ps}$ by

$$T(n, \tau)\varphi = x_n(\tau, \varphi, 0), \quad \varphi \in \mathcal{P}_{ps},$$

where $x(\cdot, \tau, \varphi, 0)$ denotes the solution of the homogeneous linear system (7.7.2) passing through (τ, φ). It is clear that the operator $T(n, \tau)$ is linear and by virtue of Axiom (PS_1) it is bounded on $\mathcal{P}_{ps}$. We denote by $\|T(n, \tau)\|_{\mathcal{P}_{ps} \to \mathcal{P}_{ps}}$ the norm of the operator $T(n, \tau)$, which satisfies the following properties:

$$T(n, s)T(s, \tau) = T(n, \tau), \quad n \geq s \geq \tau,$$
$$T(n, n) = I, \quad n \geq 0. \tag{7.7.3}$$

The operator $T(n, \tau)$ is called the solution operator of the homogeneous linear system (7.7.2) (see [148] for details).

Definition 7.7.1 ([36]). We say that (7.7.2) (or its solution operator $T(n, \tau)$) has an *exponential dichotomy*[16] on $\mathcal{P}_{ps}$ with data $(\alpha, K_{ed}, P(\cdot))$ if α, K_{ed} are positive numbers and $P(n)$, $n \in \mathbb{Z}^+$ are projections in $\mathcal{P}_{ps}$ such that, letting $Q(n) = I - P(n)$:

1. $T(n, \tau)P(\tau) = P(n)T(n, \tau), \quad n \geq \tau$.
2. The restriction $T(n, \tau)|Range(Q(\tau))$, $n \geq \tau$, is an isomorphism from $Range(Q(\tau))$ onto $Range(Q(n))$, and then we define $T(\tau, n)$ as its inverse mapping.
3. $\|T(n, \tau)\varphi\|_{\mathcal{P}_{ps}} \leq K_{ed}e^{-\alpha(n-\tau)}\|\varphi\|_{\mathcal{P}_{ps}}, \quad n \geq \tau, \quad \varphi \in P(\tau)\mathcal{P}_{ps}$.
4. $\|T(n, \tau)\varphi\|_{\mathcal{P}_{ps}} \leq K_{ed}e^{\alpha(n-\tau)}\|\varphi\|_{\mathcal{P}_{ps}}, \quad \tau > n, \quad \varphi \in Q(\tau)\mathcal{P}_{ps}$.

We assume the following condition.

Condition $(\mathcal{A}_{dif})$: $\{L(n, \cdot)\}$ is a uniformly bounded sequence of bounded linear operators mapping $\mathcal{P}_{ps}$ into $\mathbb{C}^r$. That means there is a constant $M > 1$ such that

$$|L(n, \varphi)| \leq M\|\varphi\|_{\mathcal{P}_{ps}},$$

for all $n \in \mathbb{Z}^+$ and $\varphi \in \mathcal{P}_{ps}$.

Remark 7.7.2. Condition $(\mathcal{A}_{dif})$ plays a crucial role in finding a characterization of exponential dichotomy for retarded functional difference equations in the phase space $\mathcal{P}_{ps}^\gamma$ $(\gamma > 0)$ defined by

---

[16]The problem of deciding when a functional difference equation has an *exponential dichotomy* is a priori much more complicated than for ordinary difference systems, because it is necessary to construct suitable projections; a wrong choice of projections would clearly cause very serious problems.

$$\mathcal{P}_{ps}^{\gamma} := \left\{ \varphi \colon \mathbb{Z}^{-} \to \mathbb{C}^{r} \colon \sup_{\theta \in \mathbb{Z}^{-}} \frac{|\varphi(\theta)|}{e^{-\gamma \theta}} < \infty \right\} \tag{7.7.4}$$

equipped with the norm

$$\|\varphi\|_{\mathcal{P}_{ps}^{\gamma}} = \sup_{\theta \in \mathbb{Z}^{-}} \frac{|\varphi(\theta)|}{e^{-\gamma \theta}},$$

see [36, Th. 1.1].

Theorem 7.7.3 ([3, Exponential boundedness of the solution operator]).
Assume that condition $(\mathcal{A}_{dif})$ *is fulfilled. In addition suppose that the functions*
$N(\cdot)$ *and* $M(\cdot)$ *given in Axiom* (PS_1) *are bounded.*[17] *Then there are positive*
constants $K^{\#}$ *and* $\alpha^{\#}$ *such that*

$$\|T(n,m)\|_{\mathcal{P}_{ps} \to \mathcal{P}_{ps}} \leq K^{\#} e^{\alpha^{\#}(n-m)}, \quad n \geq m \geq 0. \tag{7.7.5}$$

Proof. Without loss of generality we may assume that $M_L > \max\{1, 1/J\}$, where
J is the constant given in Axiom (PS_1). Take now

$$N^{\infty} = \max\{\|N\|_{\infty}, \|M\|_{\infty}, 1\}$$

and let $x(\cdot, m, \varphi)$ be the solution of the homogeneous system (7.7.2) passing through
(m, φ). To prove (7.7.5), we notice that in view of conditions $(\mathcal{A}_{dif})$ and (PS_1), we
have

$$\|T(m,n)\varphi\|_{\mathcal{P}_{ps}} \leq N^{\infty} \left(\sum_{j=0}^{n-m} (M_L N^{\infty})^j \right) \|\varphi\|_{\mathcal{P}_{ps}}$$

$$\leq \frac{N^{\infty}}{M_L N^{\infty} - 1} (M_L N^{\infty})^{n-m} \|\varphi\|_{\mathcal{P}_{ps}}.$$

This clearly implies (7.7.5) with

$$K^{\#} = N^{\infty}/(M_L N^{\infty} - 1) \quad \text{and} \quad \alpha^{\#} = \ln(M_L N^{\infty}).$$

$\square$

---

[17]We note that conditions of this type have been previously considered in the literature. See for
instance [58, 97, 178].

Until now there is no method to construct projections so that (7.7.2) has an *exponential dichotomy*. The next proposition shows how one can generate projections from a given one.

Proposition 7.7.4. *Under the conditions of Theorem 7.7.3, if system (7.7.2) has an exponential dichotomy with data $(\alpha, K_{ed}, P(\cdot))$, then:*

1. $\sup_{n \in \mathbb{Z}^+} \| P(n) \|_{\mathcal{P}_{ps} \to \mathcal{P}_{ps}} < \infty.$
2. $Range(P(n)) = \{ \varphi \in \mathcal{P}_{ps} : e^{-\eta(n-m)} T(n,m)\varphi \text{ is bounded for } n \geq m \}$ *for any* $0 < \eta < \alpha.$
3. *Let $\hat{P}(0)$ be a projection such that $Range(\hat{P}(0)) = Range(P(0))$. Then (7.7.2) has an exponential dichotomy on $\mathbb{Z}^+$ with data $(\alpha, \hat{K}_{ed}, \hat{P}(\cdot))$, where*

$$\hat{P}(n) = P(n) + T(n,0)\hat{P}(0)T(0,n)Q(n),$$

$$\hat{K}_{ed} = (K_{ed} + K_{ed}^2 \| \hat{P}(0) \|_{\mathcal{P}_{ps} \to \mathcal{P}_{ps}}) \sup_{m \geq 0} (1 + \| P(n) \|_{\mathcal{P}_{ps} \to \mathcal{P}_{ps}}).$$

In addition, we have

$$\sup_{m \geq 0} \| \hat{P}(m) \|_{\mathcal{P}_{ps} \to \mathcal{P}_{ps}}$$

$$\leq (1 + K_{ed}^2 \| \hat{P}(0) - P(0) \|_{\mathcal{P}_{ps} \to \mathcal{P}_{ps}}) \sup_{m \geq 0} (1 + \| P(m) \|_{\mathcal{P}_{ps} \to \mathcal{P}_{ps}}).$$

$$(7.7.6)$$

Also one has

$$\hat{P}(n) = P(n) + o(1), \quad \text{as } n \to \infty. \tag{7.7.7}$$

Proof. (1) For a fixed $\tau > 0$, set

$$\gamma_\tau = \inf\{ \| \varphi + \psi \|_{\mathcal{P}_{ps}} : \varphi \in P(\tau)\mathcal{P}_{ps}, \ \psi \in Q(\tau)\mathcal{P}_{ps}, \ \| \varphi \|_{\mathcal{P}_{ps}} = \| \psi \|_{\mathcal{P}_{ps}} = 1 \}.$$

If $\varphi \in \mathcal{P}_{ps}$ is such that $P(\tau)\varphi \neq 0$ and $Q(\tau)\varphi \neq 0$, then

$$\gamma_\tau \leq \frac{1}{\| P(\tau)\varphi \|_{\mathcal{P}_{ps}}} \left\| \varphi + \frac{\| P(n)\varphi \|_{\mathcal{P}_{ps}} - \| Q(\tau)\varphi \|_{\mathcal{P}_{ps}}}{\| Q(\tau)\varphi \|_{\mathcal{P}_{ps}}} Q(\tau)\varphi \right\| \leq \frac{2\| \varphi \|_{\mathcal{P}_{ps}}}{\| P(\tau)\varphi \|_{\mathcal{P}_{ps}}}.$$

Hence,

$$\| P(\tau) \|_{\mathcal{P}_{ps} \to \mathcal{P}_{ps}} \leq \frac{2}{\gamma_\tau}.$$

It remains to show that there is a constant $C > 0$ (independent of τ) such that $\gamma_\tau \geq C$. For this we consider $\varphi \in P(\tau)\mathcal{P}_{ps}$, $\psi \in Q(\tau)\mathcal{P}_{ps}$ with

$$\|\varphi\|_{\mathcal{P}_{ps}} = \|\psi\|_{\mathcal{P}_{ps}} = 1.$$

Taking advantage of the exponential boundedness of the solution operator, we have

$$\|\varphi + \psi\|_{\mathcal{P}_{ps}} \geq (K^{\#})^{-1} e^{-\alpha^{\#}(n-\tau)} (K_{ed}^{-1} e^{\alpha(n-\tau)} - K_{ed} e^{-\alpha(n-\tau)}) := C_{n-\tau}$$

and hence $\gamma_\tau \geq C_{n-\tau}$. Obviously, $C_m > 0$ for m sufficiently large. Thus, $0 < C_m \leq \gamma_\tau$.

(2) The inclusion

$$Range(P(m)) \subset \{\varphi \in \mathcal{P}_{ps} : e^{-n(n-m)} T(n, m)\varphi \text{ is bounded for } n \geq m\}$$

is obvious, while the converse follows from

$$\|Q(m)\varphi\|_{\mathcal{P}_{ps}} = \|T(m, n)Q(n)T(n, m)\varphi\|_{\mathcal{P}_{ps}} \leq C e^{(\alpha-\eta)(m-n)} \to 0, \quad \text{as } n \to \infty,$$

where C is a suitable constant.

(3) It is easy to see that for $n \geq \tau$

$$T(n, \tau)\hat{P}(\tau) = \hat{P}(n)T(n, \tau).$$

We recall that the operator $T(n, \tau), n \geq \tau$, is an isomorphism from $\hat{Q}(\tau)\mathcal{P}_{ps}$ to $\hat{Q}(n)\mathcal{P}_{ps}$. In fact, we define $T(\tau, n)$ as the inverse mapping which is given by

$$T(\tau, n)\hat{Q}(n)\varphi = T(\tau, n)Q(n)\varphi - T(\tau, 0)\hat{P}(0)T(0, n)Q(n)\varphi.$$

On the other hand, if $n \geq \tau$ and $\varphi \in \hat{P}(\tau)\mathcal{P}_{ps}$, then $T(n, \tau)\varphi$ is estimated by

$$\|T(n, \tau)\varphi\|_{\mathcal{P}_{ps}} \leq K_{ed} e^{-\alpha(n-\tau)} \|P(\tau)\|_{\mathcal{P}_{ps} \to \mathcal{P}_{ps}} \|\varphi\|_{\mathcal{P}_{ps}}$$
$$+ K_{ed}^2 e^{-\alpha(n-\tau)} \|\hat{P}(0)\|_{\mathcal{P}_{ps} \to \mathcal{P}_{ps}} \|Q(\tau)\|_{\mathcal{P}_{ps} \to \mathcal{P}_{ps}} \|\varphi\|_{\mathcal{P}_{ps}}$$
$$\leq \hat{K}_{ed} e^{-\alpha(n-\tau)} \|\varphi\|_{\mathcal{P}_{ps}}.$$

If $n < \tau$ and $\varphi \in \hat{Q}(\tau)\mathcal{P}_{ps}$, then we estimate $T(n, \tau)\varphi$ by

$$\|T(n, \tau)\varphi\|_{\mathcal{P}_{ps}} = e^{\alpha(n-\tau)}(K_{ed} + K_{ed}^2 \|\hat{P}(0)\|_{\mathcal{P}_{ps} \to \mathcal{P}_{ps}}) \|Q(\tau)\|_{\mathcal{P}_{ps} \to \mathcal{P}_{ps}} \|\varphi\|_{\mathcal{P}_{ps}}$$
$$\leq \hat{K}_{ed} e^{\alpha(n-\eta)} \|\varphi\|_{\mathcal{P}_{ps}}.$$

From

$$\|\hat{P}(n)-P(n)\|_{\mathcal{P}_{ps}\to\mathcal{P}_{ps}} \le K_{ed}^2 e^{-2\alpha n} \sup_{m\ge 0}(1+\|P(m)\|_{\mathcal{P}_{ps}\to\mathcal{P}_{ps}})\|\hat{P}(0)-P(0)\|_{\mathcal{P}_{ps}\to\mathcal{P}_{ps}}$$

it is easy to see that (7.7.6) and (7.7.7) hold. This completes the proof of Proposition 7.7.4. $\qquad\square$

In what follows we consider the matrix function $E^0(t)$, $t \in \mathbb{Z}^-$, defined by

$$E^0(t) = \begin{cases} I \ (r \times r \ \text{unit matrix}), & t = 0, \\ 0 \ (r \times r \ \text{zero matrix}), & t < 0. \end{cases} \tag{7.7.8}$$

We denote by $\Gamma(n, s)$ the Green function associated with (7.7.2), that is,

$$\Gamma(t, s) = \begin{cases} T(n, s+1)P(s+1) & n-1 \ge s, \\ -T(n, s+1)Q(s+1) & s > n-1. \end{cases} \tag{7.7.9}$$

For any number $1 \le p < \infty$ we shall consider the following spaces:

$$l^p(\mathbb{Z}^+;\mathcal{P}_{ps}) \equiv \{\xi:\mathbb{Z}^+ \to \mathcal{P}_{ps} \ / \ \|\xi\|_p^p := \sum_{n=0}^{\infty}\|\xi(n)\|_{\mathcal{P}_{ps}}^p < \infty\},$$

$$l^\infty(\mathbb{Z}^+;\mathcal{P}_{ps}) \equiv \{\xi:\mathbb{Z}^+ \to \mathcal{P}_{ps} \ / \ \|\xi\|_\infty := \sup_{n\in\mathbb{Z}^+}\|\xi(n)\|_{\mathcal{P}_{ps}} < \infty\}.$$

The following definition was introduced in [59].

Definition 7.7.5. We say that system (7.7.2) has *discrete maximal regularity* if for each $h \in l^p(\mathbb{Z}^+;\mathbb{C}^r)$ $(1 \le p \le \infty)$ and each $\varphi \in P(0)\mathcal{P}_{ps}$ the solution z of the boundary value problem

$$z(n+1) = L(n, z_n) + h(n), \quad n \ge 0, \tag{7.7.10}$$

$$P(0)z_0 = \varphi, \tag{7.7.11}$$

satisfies $z_\bullet \in l^p(\mathbb{Z}^+;\mathcal{P}_{ps})$.

If the system (7.7.2) has an exponential dichotomy on $\mathcal{P}_{ps}$ with data $(\alpha, K_{ed}, P(\cdot))$, then it was shown, in [59], that system (7.7.2) has discrete maximal regularity. More precisely

Theorem 7.7.6. *Assume that system (7.7.2) has an exponential dichotomy with data $(\alpha, K_{ed}, P(n))$. Then, for any $h \in \ell^p(\mathbb{Z}^+;\mathbb{C}^r)$ (with $1 \le p \le +\infty$) and any $\varphi \in \text{Range}(P(0))$, the boundary value problem (7.7.10) and (7.7.11) has a unique solution z so that $z_\bullet \in l^p(\mathbb{Z}^+;\mathcal{P}_{ps})$, namely $z = z^{sp} + z^{hom}$, where*

$$z_n^{sp} = \sum_{s=0}^{\infty} \Gamma(n,s) E^0(h(s)),$$

$$z_n^{hom} = T(n,0) P(0)\varphi.$$

This solution z satisfies $z \in \ell^{p'}(\mathbb{Z}^+; \mathbb{C}^r)$ for all $1 \leq p \leq p' \leq +\infty$, and the following estimates hold:

$$(1-e^{-\alpha})^{1-\frac{1}{p}+\frac{1}{p'}} ||z_\bullet^{sp}||_{p'} + (1-e^{-\alpha})^{1-\frac{1}{p}} ||z_0^{sp}||_{\mathcal{P}_{ps}} \leq 4K_{\mathcal{P}_{ps}} K_{ed} ||h||_p, \quad (7.7.12)$$

$$(1-e^{-\alpha})^{\frac{1}{p'}} ||z_\bullet^{hom}||_{p'} + ||z_0^{hom}||_{\mathcal{P}_{ps}} \leq 2K_{ed} ||\varphi||_{\mathcal{P}_{ps}}. \quad (7.7.13)$$

In particular, if $p = +\infty$, we get

$$(1-e^{-\alpha})(||z_\bullet^{sp}||_\infty + ||z_0^{sp}||_{\mathcal{P}_{ps}}) \leq 4K_{\mathcal{P}_{ps}} K_{ed} ||h||_\infty, \quad (7.7.14)$$

$$||z_\bullet^{hom}||_\infty + ||z_0^{hom}||_{\mathcal{P}_{ps}} \leq 2K_{ed} ||\varphi||_{\mathcal{P}_{ps}}. \quad (7.7.15)$$

Proof. First we will treat the existence problem. We observe that

$$T(n,0)z_0 + \sum_{s=0}^{n-1} T(n,s+1) E^0(h(s))$$

$$= T(n,0) P(0)\varphi - \sum_{s=0}^{\infty} T(n,s+1) Q(s+1) E^0(h(s))$$

$$+ \sum_{s=0}^{n-1} T(n,s+1) E^0(h(s))$$

$$= T(n,0) P(0)\varphi - \sum_{s=0}^{n-1} T(n,s+1) E^0(h(s))$$

$$+ \sum_{s=0}^{n-1} \Gamma(n,s) E^0(h(s)) + \sum_{s=n}^{\infty} \Gamma(n,s) E^0(h(s))$$

$$+ \sum_{s=0}^{n-1} T(n,s+1) E^0(h(s))$$

$$= z_n.$$

Hence, from [58, Lemma 2.8], we get that $z = z^{sp} + z^{hom}$ solves the boundary value problem (7.7.10) and (7.7.11). Moreover, we can infer that z is bounded. In fact, clearly $z^{hom}_\bullet$ is bounded on $\mathbb{Z}^+$. On the other hand, we have

$$||z^{sp}_\bullet||_\infty \leq 2K_{\mathcal{P}_{ps}} K_{ed} ||h||_p (1 - e^{-\alpha})^{\frac{1}{p}-1}.$$

To prove the uniqueness we use Murakami's representation formula (see [148, Theorem 2.1]) and Beyn and Lorenz's uniqueness argument in a similar manner as in [21, Theorem A.1].

We can verify that $z_\bullet \in l^p(\mathbb{Z}^+; \mathcal{P}_{ps})$. It follows from the following estimates:

$$||z^{sp}_\bullet||_p \leq 2K_{\mathcal{P}_{ps}} K_{ed} ||h||_p / (1 - e^{-\alpha}),$$

$$||z^{hom}_\bullet||_p \leq K_{ed} (1 - e^{-\alpha})^{-\frac{1}{p}} ||\varphi||_{\mathcal{P}_{ps}}.$$

Next, we will prove the estimates (7.7.12) and (7.7.13). Let p and q be conjugated exponents. We have the following estimates:

$$||z^{sp}_n||^{p'}_{\mathcal{P}_{ps}} \leq (K_{\mathcal{P}_{ps}} K_{ed})^{p'} \left(\frac{2}{1-e^{-\alpha}}\right)^{\frac{p'}{q}} \left(\sum_{s=0}^\infty e^{-\alpha|n-(s+1)|} |h(s)|^p\right)^{\frac{p'}{p}}$$

$$\leq (K_{\mathcal{P}_{ps}} K_{ed})^{p'} \left(\frac{2}{1-e^{-\alpha}}\right)^{\frac{p'}{q}} ||h||^{p'-p}_p \sum_{s=0}^\infty e^{-\alpha|n-(s+1)|} |h(s)|^p.$$

Then,

$$||z^{sp}_\bullet||^{p'}_{p'} \leq (K_{\mathcal{P}_{ps}} K_{ed})^{p'} \left(\frac{2}{1-e^{-\alpha}}\right)^{\frac{p'}{q}} ||h||^{p'-p}_p \sum_{s=0}^\infty \left(\frac{2}{1-e^{-\alpha}}\right) |h(s)|^p$$

$$= \left(2K_{\mathcal{P}_{ps}} K_{ed} (1-e^{-\alpha})^{-\frac{1}{p'}-\frac{1}{q}} ||h||_p\right)^{p'}.$$

For the second term on the left-hand side of (7.7.12) we obtain

$$||z^{sp}_0||^{p'}_{\mathcal{P}_{ps}} \leq (K_{\mathcal{P}_{ps}} K_{ed})^{p'} \left(\sum_{s=0}^\infty e^{-\alpha(s+1)} |h(s)|\right)^{p'}$$

$$\leq \left(2K_{\mathcal{P}_{ps}} K_{ed} (1-e^{-\alpha})^{-\frac{1}{q}} ||h||_p\right)^{p'}.$$

Finally, we sum

$$\|z_n^{hom}\|_{\mathcal{P}_{ps}}^{p'} \le K_{ed}^{p'} e^{-\alpha p' n} \|\varphi\|_{\mathcal{P}_{ps}}^{p'}$$

with respect to n and find

$$\|z_\bullet^{hom}\|_{p'} \le K_{ed}(1 - e^{-\alpha p'})^{-\frac{1}{p'}} \|\varphi\|_{\mathcal{P}_{ps}}$$
$$\le K_{ed}(1 - e^{-\alpha})^{-\frac{1}{p'}} \|\varphi\|_{\mathcal{P}_{ps}}.$$

This leads to the desired estimate (7.7.13). This complete the proof of Theorem 7.7.6.   □

The following result (Theorem 7.7.7) ensures the existence and uniqueness of bounded solutions of (7.7.1) which are in l^p. We need to introduce the following condition.

Condition $(\mathcal{C}_{dif})$:   The following assumptions hold:

$(\mathcal{C}_{dif}^1)$ The function $f(n, \cdot): l^p(\mathbb{Z}^+; \mathcal{P}_{ps}) \to \mathbb{C}^r$ satisfies a Lipschitz condition, that is, for all $\xi, \eta \in l^p(\mathbb{Z}^+; \mathcal{P}_{ps})$ and $n \in \mathbb{Z}^+$ we have

$$|f(n, \xi) - f(n, \eta)| \le \beta_f(n) \|\xi - \eta\|_p,$$

where $\beta_f := (\beta_f(n)) \in l^p(\mathbb{Z}^+)$;

$(\mathcal{C}_{dif}^2)$ $f(\cdot, 0) \in l^p(\mathbb{Z}^+; \mathbb{C}^r)$.

The following result is due to Agarwal et al. [3].

Theorem 7.7.7. *Assume that $(\mathcal{A}_{dif})$ is fulfilled and that the functions $N(\cdot)$ and $M(\cdot)$ given by Axiom (PS_1) are bounded. In addition assume that (7.7.2) has an exponential dichotomy on $\mathcal{P}_{ps}$ with data $(\alpha, K_{ed}, P(\cdot))$ and condition $(\mathcal{C}_{dif})$ holds. Suppose that the following condition holds:*

$$2K_{ed} K_{\mathcal{P}_{ps}} \sup_{m \in \mathbb{Z}^+} (1 + \|P(m)\|_{\mathcal{P}_{ps} \to \mathcal{P}_{ps}}) \|\beta_f\|_p + e^{-\alpha} < 1, \qquad (7.7.16)$$

where $K_{\mathcal{P}_{ps}}$ is the constant of Axiom (PS_2). Then for each $\varphi \in P(0)\mathcal{P}_{ps}$ there is a unique bounded solution y of (7.7.1) with $P(0)y_0 = \varphi$ such that $y_\bullet \in l^p(\mathbb{Z}^+; \mathcal{P}_{ps})$, in particular $y \in l^p(\mathbb{Z}^+; \mathbb{C}^r)$. Moreover, one has the following a priori estimate for the solution

$$\|y_\bullet\|_p \le C(\|\varphi\|_{\mathcal{P}_{ps}} + \|f(\cdot, 0)\|_p), \qquad (7.7.17)$$

where $C > 0$, and

$$\|y_\bullet(\varphi) - y_\bullet(\psi)\|_p$$

$$\le \frac{K_{ed}}{1 - e^{-a} - 2K_{ed} K_{\mathcal{P}_{ps}} \sup_{m \ge 0}(1 + \|P(m)\|_{\mathcal{P}_{ps} \to \mathcal{P}_{ps}}) \|\beta_f\|_p} \|\varphi - \psi\|_{\mathcal{P}_{ps}}.$$

$$(7.7.18)$$

Estimate (7.7.18) implies the continuity of the application

$$\varphi \in P(0)\mathcal{P}_{ps} \mapsto y_\bullet(\varphi) \in l^p(\mathbb{Z}^+;\mathcal{P}_{ps}).$$

Proof. Let ξ be a sequence in $l^p(\mathbb{Z}^+;\mathcal{P}_{ps})$. Using Condition $(\mathcal{C}_{dif})$ we find that the function $g(\cdot) = f(\cdot,\xi)$ is in $l^p(\mathbb{Z}^+;\mathbb{C}^r)$. In fact, we have

$$\|g\|_p^p = \sum_{n=0}^\infty |f(n,\xi)|^p$$

$$\leq \sum_{n=0}^\infty (|f(n,\xi) - f(n,0)| + |f(n,0)|)^p$$

$$\leq 2^p \sum_{n=0}^\infty |f(n,\xi) - f(n,0)|^p + 2^p \sum_{n=0}^\infty |f(n,0)|^p$$

$$\leq 2^p \sum_{n=0}^\infty \beta_f(n)^p \|\xi\|_p^p + 2^p \|f(\cdot,0)\|_p^p.$$

Hence

$$\|g\|_p \leq 2(\|\beta_f\|_p \|\xi\|_p + \|f(\cdot,0)\|_p),$$

proving that $g \in l^p(\mathbb{Z}^+;\mathcal{P}_{ps})$.

If $\varphi \in P(0)\mathcal{P}_{ps}$, by Theorem 7.7.6, system (7.7.2) has discrete maximal regularity, so that the Cauchy problem

$$\begin{cases} z(n+1) = L(n,z_n) + g(n), & n \in \mathbb{Z}^+, \\ P(0)z_0 = \varphi, \end{cases} \tag{7.7.19}$$

has a unique solution z such that $z_\bullet \in l^p(\mathbb{Z}^+;\mathcal{P}_{ps})$ which is given by

$$z_n = [\mathcal{K}\xi](n) = T(n,0)P(0)\varphi + \sum_{s=0}^\infty \Gamma(n,s)E^0(f(s,\xi)).$$

We now show that the operator

$$\mathcal{K}: l^p(\mathbb{Z}^+;\mathcal{P}_{ps}) \to l^p(\mathbb{Z}^+;\mathcal{P}_{ps})$$

has a unique fixed point.

Let ξ and η be in $l^p(\mathbb{Z}^+; \mathcal{P}_{ps})$. In view of condition $(\mathcal{C}_{dif})$ we have

$$\|\mathcal{K}\xi - \mathcal{K}\eta\|_p = \left[\sum_{n=0}^{\infty} \left\|\sum_{s=0}^{\infty} \Gamma(n,s)E^0(f(s,\xi) - f(s,\eta))\right\|_{\mathcal{P}_{ps}}^p\right]^{1/p}$$

$$\leq K_{ed}K_{\mathcal{P}_{ps}} \sup_{m\in\mathbb{Z}^+} (1 + \|P(n)\|_{\mathcal{P}_{ps}\to\mathcal{P}_{ps}})$$

$$\times \left[\sum_{n=0}^{\infty}\left(\sum_{s=0}^{\infty} e^{-\alpha|n-(s+1)|}\beta_f(s)\right)^p\right]^{1/p} \cdot \|\xi - \eta\|_p$$

$$\leq K_{ed}K_{\mathcal{P}_{ps}} \sup_{m\in\mathbb{Z}^+} (1 + \|P(m)\|_{\mathcal{P}_{ps}\to\mathcal{P}_{ps}})$$

$$\times \left[\sum_{n=0}^{\infty}\left(\frac{2}{1-e^{-\alpha}}\right)^{p/q}\left(\sum_{s=0}^{\infty} e^{-\alpha|n-(s+1)|}\beta_f(s)^p\right)\right]^{1/p} \|\xi - \eta\|_p$$

$$\leq K_{ed}K_{\mathcal{P}_{ps}} \sup_{m\in\mathbb{Z}^+} (1 + \|P(m)\|_{\mathcal{P}_{ps}\to\mathcal{P}_{ps}})\left(\frac{2}{1-e^{-\alpha}}\right)^{1/q}$$

$$\times \left[\sum_{s=0}^{\infty}\left(\frac{2}{1-e^{-\alpha}}\right)\beta_f(s)^p\right]^{1/p} \|\xi - \eta\|_p$$

$$\leq 2K_{ed}K_{\mathcal{P}_{ps}} \sup_{m\in\mathbb{Z}^+} (1 + \|P(m)\|_{\mathcal{P}_{ps}\to\mathcal{P}_{ps}})$$

$$\times (1 - e^{-\alpha})^{-1}\|\beta_f\|_p\|\xi - \eta\|_p.$$

$$(7.7.20)$$

By (7.7.16) and the contraction principle, it follows that $\mathcal{K}$ has a unique fixed point $\xi \in l^p(\mathbb{Z}^+; \mathcal{P}_{ps})$. The uniqueness of solutions is reduced to the uniqueness of the fixed point of the map $\mathcal{K}$. Indeed, let $y = y(n, 0, \varphi)$ be a solution of (7.7.1) with $P(0)y_0 = \varphi$. Considering $\xi(n) = [\mathcal{K}y_\bullet](n)$, it follows from a straightforward computation that

$$\xi(n) = T(n,0)\xi(0) + \sum_{s=0}^{n-1} T(n, s+1)E^0(f(s, y_\bullet)), \quad n \geq 0.$$

Define

$$x(n) = \begin{cases} [\xi(n)](0), & n \geq 0, \\ [\xi(0)](n), & n < 0. \end{cases}$$

Applying [58, Lemma 2.8], $x(n)$ solves the evolution equation

$$
\begin{cases}
x(n+1) = L(n, x_n) + f(n, y_\bullet), & n \geq 0, \\
P(0)x_0 = \varphi,
\end{cases}
$$

together with the relation $x_n = \xi(n)$, $n \geq 0$. Put $z_n = x_n - y_n$, so that $z(n)$ is a solution of (7.7.2) for $n \geq 0$, with $P(0)z_0 = 0$. Using Theorem 2.1 in [148], we get

$$
z_n = T(n, 0)z_0, \quad n \geq 0.
$$

Now, by property (2) of Definition 7.7.1, we find that

$$
z_0 = T(0, n)Q(n)z_n, \quad n \geq 0.
$$

By property (4) of Definition 7.7.1 and taking into account property (1) of Proposition 7.7.4, we obtain

$$
\|z_0\|_{\mathcal{P}_{ps}} \leq K_{ed} e^{-\alpha n} \sup_{m \geq 0}(1 + \|P(m)\|_{\mathcal{P}_{ps} \to \mathcal{P}_{ps}})\|z_\bullet\|_\infty, \quad n \geq 0.
$$

From this we conclude that $z_0 = 0$ and hence $z_n = 0$, which implies the uniqueness of solutions of (7.7.1).

Let ξ be the unique fixed point of $\mathcal{K}$. From condition $(\mathcal{C}_{dif})$ and (7.7.20), we have

$$
\begin{aligned}
\|\xi\|_p &= \left[\sum_{n=0}^{\infty} \left\| T(n, 0)P(0)\varphi + \sum_{s=0}^{\infty} \Gamma(n, s)E^0(f(s, \xi)) \right\|_{\mathcal{P}_{ps}}^p \right]^{1/p} \\
&\leq \left[\sum_{n=0}^{\infty} \| T(n, 0)P(0)\varphi \|_{\mathcal{P}_{ps}}^p \right]^{1/p} \\
&\quad + \left[\sum_{n=0}^{\infty} \left\| \sum_{s=0}^{\infty} \Gamma(n, s)E^0(f(s, \xi)) \right\|_{\mathcal{P}_{ps}}^p \right]^{1/p} \\
&\leq K_{ed} \left[\sum_{j=0}^{\infty} e^{-\alpha p j} \right]^{1/p} \|\varphi\|_{\mathcal{P}_{ps}} \\
&\quad + 2K_{ed}K_{\mathcal{P}_{ps}} \sup_{m \geq 0}(1 + \|P(m)\|_{\mathcal{P}_{ps} \to \mathcal{P}_{ps}})(1 - e^{-\alpha})^{-1} \\
&\quad \times \left[\sum_{s=0}^{\infty} |f(s, \xi)|^p \right]^{1/p}
\end{aligned}
$$

$$\le K_{ed}(1 - e^{-\alpha})^{-1} \|\varphi\|_{\mathcal{P}_{ps}}$$
$$+ 2K_{ed}K_{\mathcal{P}_{ps}} \sup_{m \ge 0}(1 + \|P(m)\|_{\mathcal{P}_{ps} \to \mathcal{P}_{ps}})(1 - e^{-\alpha})^{-1}$$
$$\times (\|\beta_f\|_p \|\xi\|_p + \|f(\cdot, 0)\|_p),$$

whence

$$\|\xi\|_p \le \frac{K_{ed} \max\{1, 2K_{\mathcal{P}_{ps}}\} \sup_{m \ge 0}(1 + \|P(m)\|_{\mathcal{P}_{ps} \to \mathcal{P}_{ps}})}{1 - e^{-\alpha} - 2K_{ed}K_{\mathcal{P}_{ps}} \sup_{m \ge 0}(1 + \|P(m)\|_{\mathcal{P}_{ps} \to \mathcal{P}_{ps}})\|\beta_f\|_p}$$
$$\times \left[\|\varphi\|_{\mathcal{P}_{ps}} + \|f(\cdot, 0)\|_p \right].$$

Our next step is to show (7.7.18), and for this we argue as follows:

$$\|y_\bullet(\varphi) - y_\bullet(\psi)\|_p$$
$$\le K_{ed}(1 - e^{-\alpha})^{-1} \|\varphi - \psi\|_{\mathcal{P}_{ps}}$$
$$+ 2K_{ed}K_{\mathcal{P}_{ps}} \sup_{m \ge 0}(1 + \|P(m)\|_{\mathcal{P}_{ps} \to \mathcal{P}_{ps}})(1 - e^{-\alpha})^{-1}$$
$$\times \left[\sum_{s=0}^{\infty} |f(s, y_\bullet(\varphi)) - f(s, y_\bullet(\psi))|^p \right]^{1/p}$$
$$\le K_{ed}(1 - e^{-\alpha})^{-1} \|\varphi - \psi\|_{\mathcal{P}_{ps}}$$
$$+ 2K_{ed}K_{\mathcal{P}_{ps}} \sup_{m \ge 0}(1 + \|P(m)\|_{\mathcal{P}_{ps} \to \mathcal{P}_{ps}})(1 - \alpha^{-\alpha})^{-1}$$
$$\times \|\beta_f\|_p \|y_\bullet(\varphi) - y_\bullet(\psi)\|_p.$$

It is easy to see that the desired bounds (7.7.18) follow from the above inequality. This finishes the proof of the theorem.   □

Example 7.7.8. Assume that $(\mathcal{A}_{dif})$ is fulfilled for the phase space $\mathcal{P}_{ps}^{\gamma}$ [see (7.7.4)] and suppose that (7.7.2) has an exponential dichotomy on $\mathcal{P}_{ps}^{\gamma}$. Let $\{\tilde{\mathcal{L}}(n, \cdot)\}_{n \in \mathbb{Z}^+}$ be a sequence of bounded linear operators from $\mathcal{P}_{ps}^{\gamma}$ into $\mathbb{C}^r$. If $\sup_{n \in \mathbb{Z}^+} \|\tilde{\mathcal{L}}(n, \cdot)\|_{\mathcal{P}_{ps}^{\gamma} \to \mathbb{C}^r}$ is sufficiently small, then by the robustness of the exponential dichotomy (see Theorem 7.10.2) we find that equation

$$x(n + 1) = L(n, x_n) + \tilde{\mathcal{L}}(n, x_n), \quad n \ge 0,$$

has an exponential dichotomy (on $\mathcal{P}_{ps}^{\gamma}$) as well for suitable data $(\alpha_*, K_*, P_*(\cdot))$.
    Next, assume that condition $(\mathcal{C}_{dif})$ holds and that

$$2K_* \sup_{m \in \mathbb{Z}^+} (1 + \|P_*(m)\|_{\mathcal{P}_{ps}^\gamma \to \mathcal{P}_{ps}^\gamma}) \|\beta_f\|_p + e^{-\alpha_*} < 1.$$

Then by Theorem 7.7.7 for each $\varphi \in P_*(0)\mathcal{B}_{ps}^\gamma$, there is a unique bounded solution x of the initial value problem

$$\begin{cases} x(n+1) = L(n, x_n) + \tilde{L}(n, x_n) + f(n, x_\bullet), & n \in \mathbb{Z}^+, \\ P_*(0)x_0 = \varphi, \end{cases}$$

such that $x_\bullet \in l^p(\mathbb{Z}^+; \mathcal{P}_{ps}^\gamma)$ and in particular $x \in l^p(\mathbb{Z}^+; \mathbb{C}^r)$. Moreover, we have the following a priori estimate for the solution:

$$\|x_\bullet\|_p \le C(\|\varphi\|_{\mathcal{P}_{ps}^\gamma} + \|f(\cdot, 0)\|_p),$$

where $C > 0$. This finishes the discussion of Example 7.7.8.

Next we will establish a version of Theorem 7.7.7 which enables us to consider locally Lipschitz perturbations of (7.7.1).

To state the next result we need to introduce the following assumption.

Condition $(\mathcal{D}_{dif})$:   Suppose that the following conditions hold:
$(\mathcal{D}_{dif}^1)$ The function $f: \mathbb{Z}^+ \times l^p(\mathbb{Z}^+; \mathcal{P}_{ps}) \to \mathbb{C}^r$ is locally Lipschitz with respect
to the second variable, that is, for each positive number R, for all $n \in \mathbb{Z}^+$
and for all $\xi, \eta \in l^p(\mathbb{Z}^+; \mathcal{P}_{ps})$ with $\|\xi\|_p \le R$, $\|\eta\|_p \le R$,

$$|f(n, \xi) - f(n, \eta)| \le \mathfrak{l}(n, R)\|\xi - \eta\|_p$$

where $\mathfrak{l}: \mathbb{Z}^+ \times [0, \infty) \to [0, \infty)$ is a nondecreasing function with respect
to the second variable.
$(\mathcal{D}_{dif}^2)$ There is a positive number a such that $\sum_{n=0}^\infty \mathfrak{l}(n, a)^p < \infty$.
$(\mathcal{D}_{dif}^3)$ $f(\cdot, 0) \in l^p(\mathbb{Z}^+; \mathbb{C}^r)$.

We introduce some basic notations. We denote by $\mathcal{L}_m^p = \mathcal{L}_m^p(\mathbb{Z}^+; \mathcal{P}_{ps})$ the closed subspace of $l^p(\mathbb{Z}^+; \mathcal{P}_{ps})$ of the sequences $\xi = (\xi(n))$ such that $\xi(n) = 0$ if $0 \le n \le m$. For $\lambda > 0$, denote by $\mathcal{L}_m^p[\lambda]$ the ball $\|\xi\|_p \le \lambda$ in $\mathcal{L}_m^p$.

The following result is due to Agarwal et al. [3].

Theorem 7.7.9. *Assume that conditions $(\mathcal{A}_{dif})$ and $(\mathcal{D}_{dif})$ are fulfilled and that the functions $N(\cdot)$ and $M(\cdot)$ given by Axiom (PS_1) are bounded. In addition suppose that (7.7.2) has an exponential dichotomy on $\mathcal{P}_{ps}$ with data $(\alpha, K_{ed}, P(\cdot))$. Then there are positive constants $M^\circ \in \mathbb{R}$ and $m \in \mathbb{Z}^+$ such that for each $\varphi \in P(m)\mathcal{P}_{ps}$ with $\|\varphi\|_{\mathcal{P}_{ps}} \le M^\circ$, there is a unique bounded solution y of (7.7.1) for $n \ge m$ with $P(m)y_m = \varphi$ such that $y_n = 0$ for $0 \le n \le m$ and $\|y_\bullet\|_p \le a$, where a is the constant of condition $(\mathcal{D}_{dif}^2)$. In particular, $y_n = o(1)$ as $n \to \infty$.*

Proof. Let $v \in (0, 1)$. Using $(\mathcal{D}_{dif}^2)$ and $(\mathcal{D}_{dif}^3)$ there are n_1 and n_2 in $\mathbb{Z}^+$ such that

$$\frac{2K_{\mathcal{P}_{ps}} K_{ed} \sup_{n \in \mathbb{Z}^+} (1 + \|P(n)\|_{\mathcal{P}_{ps} \to \mathcal{P}_{ps}})}{1 - e^{-\alpha}} \left[\sum_{j=n_1}^{\infty} |f(j, 0)| \right]^{1/p} \leq \frac{v}{2} a \quad (7.7.21)$$

and

$$\tau := v + \frac{2K_{\mathcal{P}_{ps}} K_{ed} \sup_{n \in \mathbb{Z}^+} (1 + \|P(n)\|_{\mathcal{P}_{ps} \to \mathcal{P}_{ps}})}{1 - e^{-\alpha}} \left[\sum_{j=n_2}^{\infty} |f(j, 0)| \right]^{1/p}$$

$$< 1, \quad (7.7.22)$$

where $K_{\mathcal{P}_{ps}}$ is the constant of Axiom (PS_2). Let us denote

$$m := \max\{n_1, n_2\},$$

$$M^{\odot} := \frac{va(1 - e^{-\alpha})}{2K_{\mathcal{P}_{ps}} K_{ed} \sup_{n \in \mathbb{Z}^+} (1 + \|P(n)\|_{\mathcal{P}_{ps} \to \mathcal{P}_{ps}})}$$

and let $\varphi \in P(m)\mathcal{P}_{ps}$ be such that $\|\varphi\|_{\mathcal{P}_{ps}} \leq M^{\odot}$. Let ξ be a sequence in $\mathcal{L}_{m-1}^p[a]$. A short argument involving condition $(\mathcal{D}_{dif})$ shows that the sequence

$$g_n := \begin{cases} 0, & 0 \leq n < m \\ f(n, \xi), & n \geq m, \end{cases} \quad (7.7.23)$$

belongs to $l^p(\mathbb{Z}^+; \mathbb{C}^r)$. By discrete maximal regularity (see Theorem 7.7.6) the Cauchy problem

$$\begin{cases} z(n + 1) = L(n, z_n) + g(n), & n \in \mathbb{Z}^+, \\ P(m)z_m = \varphi, \end{cases} \quad (7.7.24)$$

has a unique solution z such that

$$\sum_{n=m}^{\infty} \|z_n\|_{\mathcal{P}_{ps}}^p < \infty,$$

which is given by

$$z_n = [\mathfrak{F}\xi](n) = T(n, m)P(m)\varphi + \sum_{s=m}^{\infty} \Gamma(n, s)E^0(f(s, \xi)), \quad n \geq m. \quad (7.7.25)$$

We define $[\mathfrak{F}\xi](n) = 0$ if $0 \leq n < m$.

Taking into account condition $(\mathcal{D}_{dif})$, we have the following estimates which imply that $\mathfrak{F}\xi$ belongs to $\mathcal{L}^p_{m-1}[a]$:

$$\left[\sum_{n=m}^{\infty} \|T(n,m)P(m)\varphi\|^p_{\mathcal{P}_{ps}}\right]^{1/p}$$

$$\leq K_{ed} \sup_{m\in\mathbb{Z}^+}(1 + \|P(m)\|_{\mathcal{P}_{ps}\to\mathcal{P}_{ps}})\left[\sum_{j=0}^{\infty} e^{-\alpha pj}\right]^{1/p}\|\varphi\|_{\mathcal{P}_{ps}} \qquad (7.7.26)$$

$$\leq K_{ed} \sup_{m\in\mathbb{Z}^+}(1 + \|P(m)\|_{\mathcal{P}_{ps}\to\mathcal{P}_{ps}})(1 - e^{-\alpha})^{-1}\|\varphi\|_{\mathcal{P}_{ps}},$$

$$\left[\sum_{n=m}^{\infty}\left\|\sum_{s=m}^{\infty}\Gamma(n,s)E^0(f(s,0))\right\|^p_{\mathcal{P}_{ps}}\right]^{1/p}$$

$$\leq K_{ed}K_{\mathcal{P}_{ps}} \sup_{m\in\mathbb{Z}^+}(1 + \|P(m)\|_{\mathcal{P}_{ps}\to\mathcal{P}_{ps}})$$

$$\times\left[\sum_{n=m}^{\infty}\left(\sum_{s=m}^{\infty} e^{-\alpha|n-(s+1)|}|f(s,0)|\right)^p\right]^{1/p}$$

$$\leq K_{ed}K_{\mathcal{P}_{ps}} \sup_{m\in\mathbb{Z}^+}(1 + \|P(m)\|_{\mathcal{P}_{ps}\to\mathcal{P}_{ps}})\left(\frac{2}{1 - e^{-\alpha}}\right)^{1/q}$$

$$\times\left[\sum_{n=m}^{\infty}\sum_{s=m}^{\infty} e^{-\alpha|n-(s+1)|}|f(s,0)|^p\right]^{1/p} \qquad (7.7.27)$$

$$\leq 2K_{ed}K_{\mathcal{P}_{ps}} \sup_{m\in\mathbb{Z}^+}(1 + \|P(m)\|_{\mathcal{P}_{ps}\to\mathcal{P}_{ps}})$$

$$\times (1 - e^{-\alpha})^{-1}\left[\sum_{s=m}^{\infty}|f(s,0)|^p\right]^{1/p}.$$

Analogously we have

$$\left[\sum_{n=m}^{\infty}\left\|\sum_{s=m}^{\infty}\Gamma(n,s)E^0(f(s,\xi) - f(s,0))\right\|^p_{\mathcal{P}_{ps}}\right]^{1/p}$$

$$\leq 2aK_{ed}K_{\mathcal{P}_{ps}} \sup_{m\in\mathbb{Z}^+} (1 + \|P(m)\|_{\mathcal{P}_{ps}\to\mathcal{P}_{ps}})(1 - e^{-\alpha})^{-1}\left[\sum_{s=m}^{\infty} ł(s,a)^p\right]^{1/p}.$$

$$(7.7.28)$$

Then, inequalities (7.7.26)–(7.7.28) together with (7.7.21) and (7.7.22) imply

$$\|\mathfrak{F}\xi\|_p \leq \frac{K_{ed}}{1 - e^{-\alpha}} \sup_{m\in\mathbb{Z}^+} (1 + \|P(m)\|_{\mathcal{P}_{ps}\to\mathcal{P}_{ps}})$$

$$\times \frac{va(1 - e^{-\alpha})}{2K_{ed}\sup_{m\in\mathbb{Z}^+}(1 + \|P(m)\|_{\mathcal{P}_{ps}\to\mathcal{P}_{ps}})}$$

$$+\frac{v}{2}a + (\tau - v)a$$

$$= \tau a \leq a, \qquad\qquad (7.7.29)$$

proving that $\mathfrak{F}\xi$ belongs to $\mathcal{L}^p_{m-1}[a]$. On the other hand, for all ξ and η in $\mathcal{L}^p_{m-1}[a]$, we obtain

$$\|\mathfrak{F}\xi - \mathfrak{F}\eta\|_p = \left[\sum_{n=m}^{\infty}\left\|\sum_{s=m}^{\infty} \Gamma(n,s)E^0\big(f(s,\xi) - f(s,\eta)\big)\right\|_{\mathcal{P}_{ps}}^p\right]^{1/p}$$

$$\leq 2K_{ed}K_{\mathcal{P}_{ps}} \sup_{m\in\mathbb{Z}^+} (1 + \|P(m)\|_{\mathcal{P}_{ps}\to\mathcal{P}_{ps}})$$

$$\times (1 - e^{-\alpha})^{-1}\left[\sum_{s=m}^{\infty} ł(s,a)^p\right]^{1/p}\|\xi - \eta\|_p$$

$$\leq (\tau - v)\|\xi - \eta\|_p.$$

Hence $\mathfrak{F}$ is a $(\tau - v)$-contraction on $\mathcal{L}^p_{m-1}[a]$.

Next we will establish the uniqueness of solutions. Let $y = y(n, m, \psi)$ be a solution of (7.7.1) with the properties stated there. Considering $z(n) = [\mathfrak{F}y_\bullet](n)$, it follows from a straightforward computation that

$$z(n) = T(n,m)z(m) + \sum_{s=m}^{n-1} T(n, s + 1)E^0(f(s, y_\bullet)), \quad n \geq m.$$

We define a function $a: \mathbb{Z} \to \mathbb{C}^r$ by

$$a(n) = \begin{cases} [z(n)](m), & n \geq m, \\ [z(m)](n - m), & n < m. \end{cases}$$

Applying [58, Lemma 2.8], we can infer that $a(n)$ satisfies $a_n = z(n)$, $n \geq m$, and it is a solution of

$$\begin{cases} a(n+1) = L(n, a_n) + f(y, y_\bullet), & n \geq m, \\ P(m)a_m = \varphi. \end{cases}$$

Then the difference $x_n = a_n - y_n$ is a solution of the linear problem (7.7.2) for $n \geq m$. In this way, we have

$$y_n = [\mathfrak{F}y_\bullet](n) + T(n, m)([\mathfrak{F}y_\bullet](m) - \psi).$$

Now, putting

$$\Phi(n) := \left\| T(n, m)Q(m)([\mathfrak{F}y_\bullet](m) - \psi) \right\|_{\mathcal{P}_{ps}}^{-1}$$

and $\Psi(n) := \sum_{s=n}^{\infty} \Phi(s+1)$, we have

$$\frac{\Psi(n)}{\Phi(n)} = \sum_{s=n}^{\infty} \left\| T(n, m)Q(m)([\mathfrak{F}y_\bullet](m) - \psi) \right\|_{\mathcal{P}_{ps}} \Phi(s+1)$$

$$\leq K_{ed} \sum_{s=n}^{\infty} e^{\alpha(n-(s+1))} \|Q(s+1)\|_{\mathcal{P}_{ps} \to \mathcal{P}_{ps}} \Phi(s+1)^{-1} \Phi(s+1)$$

$$\leq K_{ed} \sup_{m \in \mathbb{Z}^+} (1 + \|P(m)\|_{\mathcal{P}_{ps} \to \mathcal{P}_{ps}})(1 - e^{-\alpha})^{-1},$$

hence

$$\Psi(n) \leq K_{ed} \sup_{m \in \mathbb{Z}^+} (1 + \|P(m)\|_{\mathcal{P}_{ps} \to \mathcal{P}_{ps}})(1 - e^{-\alpha})^{-1} \Phi(n),$$

for all $n \geq m$, which means that

$$\left\{ \left\| T(n, m)Q(m)([\mathfrak{F}y_\bullet](m) - \psi) \right\|_{\mathcal{P}_{ps}} \right\}_{n \geq m}$$

is unbounded. Since

$$T(n, m)([\mathfrak{F}y_\bullet](m) - \psi) = T(n, m)Q(m)([\mathfrak{F}y_\bullet](m) - \psi)$$

and $y_\bullet$, $\mathfrak{F}y_\bullet$ are bounded, we conclude that $y_n = [\mathfrak{F}y_\bullet](n)$. Hence the uniqueness of y follows from the uniqueness of the fixed point of the map $\mathfrak{F}$. This completes the proof of the theorem. $\qquad\square$

7.8 More About Boundedness and Asymptotic Behavior for RFDE

We shall consider the following spaces:

$$l_\varrho^\infty(\mathbb{Z}^+; \mathcal{P}_{ps}) := \{\xi : \mathbb{Z}^+ \to \mathcal{P}_{ps} \mathbin{/} \|\xi\|_\varrho = \sup_{n \in \mathbb{Z}^+} \|\xi(n)\|_{\mathcal{P}_{ps}} e^{-\varrho n} < \infty\}, \quad (7.8.1)$$

$$l_\varrho^1(\mathbb{Z}^+; \mathbb{C}^r) := \{\varphi : \mathbb{Z}^+ \to \mathbb{C}^r \mathbin{/} \|\varphi\|_{1,\varrho} := \sum_{n=0}^\infty |\varphi(n)| e^{-\varrho n} < \infty\}, \quad (7.8.2)$$

where ϱ is a positive constant. To state the next result we introduce the following condition.

Condition $(\mathcal{E}_{dif})$:   Suppose that the following statements hold:
$(\mathcal{E}_{dif}^1)$ The function $g(n, \cdot) : \mathcal{P}_{ps} \to \mathbb{C}^r$ satisfies a Lipschitz condition for all $n \in \mathbb{Z}^+$, that is, for all $\varphi, \psi \in \mathcal{P}_{ps}$ and $n \in \mathbb{Z}^+$, we have

$$|g(n, \varphi) - g(n, \psi)| \le l_g(n) \|\varphi - \psi\|_{\mathcal{P}_{ps}},$$

where $l_g := (l_g(n)) \in l_1(\mathbb{Z}^+; \mathbb{R}^+)$.
$(\mathcal{E}_{dif}^2)$ $g(\cdot, 0) \in l_{\alpha^\#}^1(\mathbb{Z}^+; \mathbb{C}^r)$, where $\alpha^\#$ is given in Theorem 7.7.3.
   We have the following result about weighted bounded solutions due to Agarwal et al. [3].

Theorem 7.8.1. *Assume that conditions $(\mathcal{A}_{dif})$ and $(\mathcal{E}_{dif})$ hold and that the functions $N(\cdot)$ and $M(\cdot)$ given by Axiom (PS_1) are bounded. Let $K^\#$ and $\alpha^\#$ be the constants of Theorem 7.7.3, then there is a unique weighted bounded solution y of the evolution equation*

$$y(n + 1) = L(n, y_n) + g(n, y_n), \quad (7.8.3)$$

for $n \ge 0$ with $y_0 = 0$ and $y_\bullet \in l_{\alpha^\#}^\infty$. Moreover, we have the following a priori estimate for the solution:

$$\|y_\bullet\|_{\alpha^\#} \le K^\# K_{\mathcal{P}_{ps}} e^{-\alpha^\#} \|g(\cdot, 0)\|_{1,\alpha^\#} e^{K^\# K_{\mathcal{P}_{ps}} e^{-\alpha^\#} \|l_g\|_1}, \quad (7.8.4)$$

where $y_\bullet$ denotes the $\mathcal{P}_{ps}$-valued function defined by $n \longrightarrow y_n$, and $y_n : \mathbb{Z}^- \longrightarrow \mathbb{C}^r$ is the history function.

Proof. We define the operator Ω on $l_{\alpha^\#}^\infty$ by

$$[\Omega \xi](n) = \sum_{s=0}^{n-1} T(n, s + 1) E^0(g(s, \xi(s))), \quad \xi \in l_{\alpha^\#}^\infty. \quad (7.8.5)$$

We now show that the operator $\Omega: l^\infty_{\alpha^\#} \to l^\infty_{\alpha^\#}$ has a unique fixed point. We observe that Ω is well defined. In fact, we use condition $(\mathcal{E}_{dif})$ and Theorem 7.7.3 to obtain

$$\|[\Omega\xi](n)\|_{\mathcal{P}_{ps}} e^{-\alpha^\# n} \le K^\# K_{\mathcal{P}_{ps}} e^{-\alpha^\#} \sum_{s=0}^{n-1} |g(s,\xi(s))| e^{-\alpha^\# s}$$

$$\le K^\# K_{\mathcal{P}_{ps}} e^{-\alpha^\#} \left[\sum_{s=0}^{n-1} l_g(s) \|\xi(s)\|_{\mathcal{P}_{ps}} e^{-\alpha^\# s} + \sum_{s=0}^{n-1} |g(s,0)| e^{-\alpha^\# s} \right]$$

whence

$$\|\Omega\xi\|_{\alpha^\#} \le K^\# K_{\mathcal{P}_{ps}} e^{-\alpha^\#} \left[\|l_g\|_1 \|\xi\|_{\alpha^\#} + \|g(\cdot,0)\|_{1,\alpha^\#} \right]. \qquad (7.8.6)$$

It proves that the space $l^\infty_{\alpha^\#}$ is invariant under Ω.

Let ξ and η be in $l^\infty_{\alpha^\#}$. In view of Theorem 7.7.3 and condition $(\mathcal{E}^1_{dif})$, we have initially

$$\left\|[\Omega\xi](n) - [\Omega\eta](n)\right\|_{\mathcal{P}_{ps}} e^{-\alpha^\# n} \le K^\# K_{\mathcal{P}_{ps}} e^{-\alpha^\#} \left(\sum_{s=0}^{n-1} l_g(s) \right) \|\xi - \eta\|_{\alpha^\#}$$

$$\le K^\# K_{\mathcal{P}_{ps}} e^{-\alpha^\#} \|l_g\|_1 \|\xi - \eta\|_{\alpha^\#}. \qquad (7.8.7)$$

Next we consider the iterates of the operator Ω. Taking into account Proposition 4.3.1, we observe from (7.8.7) that

$$\left\|[\Omega^2\xi](n) - [\Omega^2\eta](n)\right\|_{\mathcal{P}_{ps}} e^{-\alpha^\# n}$$

$$\le K^\# K_{\mathcal{P}_{ps}} e^{-\alpha^\#} \sum_{s=0}^{n-1} l_g(s) \left\|[\Omega\xi](s) - [\Omega\eta](s)\right\|_{\mathcal{P}_{ps}} e^{-\alpha^\# s}$$

$$\le [K^\# K_{\mathcal{P}_{ps}} e^{-\alpha^\#}]^2 \left(\sum_{s=0}^{n-1} l_g(s) \left(\sum_{i=0}^{s-1} l_g(i) \right) \right) \|\xi - \eta\|_{\alpha^\#}$$

$$\le \frac{1}{2} [K^\# K_{\mathcal{P}_{ps}} e^{-\alpha^\#}]^2 \left(\sum_{s=0}^{n-1} l_g(s) \right)^2 \|\xi - \eta\|_{\alpha^\#}.$$

Therefore,

$$\|[\Omega^2\xi](n) - [\Omega^2\eta](n)\|_{\alpha^\#} \le \frac{1}{2} [K^\# K_{\mathcal{P}_{ps}} e^{-\alpha^\#} \|l_g\|_1]^2 \|\xi - \eta\|_{\alpha^\#}.$$

In general, proceeding by induction, we can assert that

$$||[\Omega^n \xi](n) - [\Omega^n \eta](n)||_{\alpha^\#} \leq \frac{1}{n!}[K^\# K_{\mathcal{P}_{ps}} e^{-\alpha^\#}||l_g||_1]^n ||\xi - \eta||_{\alpha^\#}.$$

The above estimates imply that the operator Ω^n is a contraction for n sufficiently large. Therefore, Ω has a unique fixed point in $l_{\alpha^\#}^\infty$. The uniqueness of solutions of (7.8.3) is reduced to the uniqueness of the fixed point of the map Ω.

Let ξ be the unique fixed point of Ω. Then by condition $(\mathcal{E}_{dif})$ we have

$$||\xi(n)||_{\mathcal{P}_{ps}} e^{-\alpha^\# n} \leq K^\# K_{\mathcal{P}_{ps}} e^{-\alpha^\#} \left(||g(\cdot, 0)||_{1,\alpha^\#} + \sum_{s=0}^{n-1} l_g(s) ||\xi(s)||_{\mathcal{P}_{ps}} e^{-\alpha^\# s} \right).$$

Then as an application of the discrete Gronwall inequality we get

$$||\xi||_{\alpha^\#} \leq K^\# K_{\mathcal{P}_{ps}} e^{-\alpha^\#} ||g(\cdot, 0)||_{1,\alpha^\#} e^{K^\# K_{\mathcal{P}_{ps}} e^{-\alpha^\#} ||l_g||_1}.$$

This finishes the proof of the theorem.   □

Next, we are concerned with the initial value problem defined by the semilinear difference equation with infinite delay

$$x(n + 1) = \mathcal{F}(x_n) + g(n, x_n), \quad n \geq 0, \tag{7.8.8}$$

with the initial condition

$$x_0 = \varphi \in \mathcal{P}_{ps}, \tag{7.8.9}$$

where $\mathcal{F}: \mathcal{P}_{ps} \to \mathbb{C}^r$ is a bounded operator and $g: \mathbb{Z}^+ \times \mathcal{P}_{ps} \to \mathbb{C}^r$.

The following result ensures the existence and uniqueness of bounded solutions of problem (7.8.8) and (7.8.9). To study this initial value problem we assume the following condition.

Condition $(\mathcal{G}_{dif})$:   The solution operator $T(n)$ of equation

$$x(n + 1) = \mathcal{F}(x_n), \quad n \in \mathbb{Z}^+, \tag{7.8.10}$$

decays exponentially, that is, there are constants $K_{exp} \geq 1$ and $\alpha^b > 0$ such that

$$||T(n)||_{\mathcal{P}_{ps} \to \mathcal{P}_{ps}} \leq K_{exp} e^{-\alpha^b n}, \quad n \in \mathbb{Z}^+. \tag{7.8.11}$$

Corollary 7.8.2. *Assume that the solution operator $T(n)$ of (7.8.10) decays exponentially. In addition, suppose that condition $(\mathcal{E}_{dif}^1)$ holds with $g(\cdot, 0) \in l_{\alpha^b}^1(\mathbb{Z}^+; \mathbb{C}^r)$, where α^b is the constant in (7.8.11). Then there is a unique weighted bounded solution y of (7.8.8) with $y_0 = 0$ such that*

$$||y_\bullet||_{\alpha^b} = \sup_{n\in\mathbb{Z}^+} ||y_n||_{\mathcal{P}_{ps}} e^{-\alpha^b n} < \infty.$$

Moreover, we have the a priori estimate

$$||y_\bullet||_{\alpha^b} \le K_{exp} K_{\mathcal{P}_{ps}} e^{-\alpha^b} ||g(\cdot,0)||_{1,\alpha} e^{K_{exp} K_{\mathcal{P}_{ps}} e^{-\alpha^b} ||l_g||_1}, \qquad (7.8.12)$$

where K_{exp} and $K_{\mathcal{P}_{ps}}$ are the constants given by (7.8.11) and Axiom (PS_2) respectively.

Now we establish a local version of the Theorem 7.8.1. To state such version, we will require the following assumption.

Condition $(\mathcal{H}_{dif})$: Suppose that the following conditions hold:

$(\mathcal{H}_{dif}^1)$ The function $g: \mathbb{Z}^+ \times \mathcal{P}_{ps} \to \mathbb{C}^r$ is locally Lipschitz with respect to the second variable, that is, for each positive number R, for all $n \in \mathbb{Z}^+$ and for all $\varphi, \psi \in \mathcal{P}_{ps}$ with $||\varphi||_{\mathcal{P}_{ps}} \le R$, $||\psi||_{\mathcal{P}_{ps}} \le R$, we have

$$|g(n,\varphi) - g(n,\psi)| \le \rho(n,R)||\varphi - \psi||_{\mathcal{P}_{ps}},$$

where $\rho: \mathbb{Z}^+ \times [0,\infty) \to [0,\infty)$ is a nondecreasing function with respect to the second variable.

$(\mathcal{H}_{dif}^2)$ There is a positive number a such that

$$\sum_{n=0}^{\infty} \rho(n, ae^{\alpha^\# n}) < \infty,$$

where $\alpha^\#$ is the constant given in Theorem 7.7.3.

$(\mathcal{H}_{dif}^3)$ $g(\cdot,0) \in l_{\alpha^\#}^1(\mathbb{Z}^+; \mathbb{C}^r)$.

Next we have the following result.

Theorem 7.8.3. *Assume that conditions $(\mathcal{A}_{dif})$ and $(\mathcal{H}_{dif})$ are fulfilled and that the functions $N(\cdot)$ and $M(\cdot)$ given by Axiom (PS_1) are bounded. Then there is a unique weighted bounded solution y of (7.8.3) for $n > m$ with $y_m = 0$ such that $\sup_{n>m} ||y_n||_{\mathcal{P}_{ps}} e^{-\alpha^\# n} \le a$, where a is the constant of condition $(\mathcal{H}_{dif}^2)$.*

Proof. Let $\nu \in (0,1)$. Using $(\mathcal{H}_{dif}^2)$ and $(\mathcal{H}_{dif}^3)$ there are n_1 and n_2 in $\mathbb{Z}^+$ such that

$$K_{\mathcal{P}_{ps}} K^\# e^{-\alpha^\#} \sum_{s=n_1}^{\infty} |g(s,0)| e^{-\alpha^\# s} \le \nu a, \qquad (7.8.13)$$

and

$$\tau := \nu + K_{\mathcal{P}_{ps}} K^\# e^{-\alpha^\#} \sum_{s=n_2}^{\infty} \mathfrak{t}(s, ae^{-\alpha^\# s}) < 1, \qquad (7.8.14)$$

where $K^\#$ and $\alpha^\#$ are the constants of Theorem 7.7.3. Set $m = \max\{n_1, n_2\}$. Denote by $\mathcal{L}_\#^\infty$ the Banach space of all weighted bounded functions $\xi : \mathbb{N}(m + 1) \to \mathcal{P}_{ps}$ equipped with the norm

$$\|\xi\|_{\alpha^\#} = \sup_{n \ge m+1} \|\xi(n)\|_{\mathcal{P}_{ps}} e^{-\alpha^\# n}.$$

Denote by $\mathcal{L}_\#^\infty[a]$ the ball $\|\xi\|_{\alpha^\#} \le a$ in $\mathcal{L}_\#^\infty$.
   We define the operator Ω on $\mathcal{L}_\#^\infty[a]$ by

$$[\Omega\xi](n) = \sum_{s=m}^{n-1} T(n, s + 1) E^0(g(s, \xi(s))). \qquad (7.8.15)$$

We observe that Ω is well defined. In fact,

$$\|[\Omega\xi](n)\|_{\mathcal{P}_{ps}} e^{-\alpha^\# n}$$

$$\le K_{\mathcal{P}_{ps}} K^\# e^{-\alpha^\#} \sum_{s=m}^{\infty} e^{-\alpha^\# s} |g(s, \xi(s))|$$

$$\le K_{\mathcal{P}_{ps}} K^\# e^{-\alpha^\#} \sum_{s=m}^{n-1} e^{-\alpha^\# s} \left(\mathfrak{t}(s, ae^{-\alpha^\# s}) \|\xi(s)\|_{\mathcal{P}_{ps}} + |g(s, 0)|\right)$$

$$\le (\tau - \nu)a + \nu a \le \tau a < a,$$

whence $\|\Omega\xi\|_{\alpha^\#} \le a$.
   Let ξ and η be in $\mathcal{L}_\#^\infty[a]$. We have

$$\|[\Omega\xi](n) - [\Omega\eta](n)\|_{\mathcal{P}_{ps}} e^{-\alpha^\# n}$$

$$\le K_{\mathcal{P}_{ps}} K^\# e^{-\alpha^\#} \sum_{s=m}^{n-1} \mathfrak{t}(s, ae^{-\alpha^\# s}) \|\xi(s) - \eta(s)\|_{\mathcal{P}_{ps}} e^{-\alpha^\# s}$$

$$\le K_{\mathcal{P}_{ps}} K^\# e^{-\alpha^\#} \sum_{s=m}^{\infty} \mathfrak{t}(s, ae^{-\alpha^\# s}) \|\xi - \eta\|_{\alpha^\#}$$

$$\le (\tau - \nu) \|\xi - \eta\|_{\alpha^\#}.$$

Hence Ω is a $(\tau - \nu)$-contraction. This completes the proof of the theorem. $\square$

The exponential dichotomy gives us relevant information about the asymptotic relation between weighted bounded solutions of (7.7.2) and its perturbed system (7.8.3). Precisely we have the following theorem on asymptotic.

Theorem 7.8.4 ([3]). *Assume that condition* $(\mathcal{A}_{dif})$ *holds and that the functions* $N(\cdot)$ *and* $M(\cdot)$ *given by Axiom* (PS_1) *are bounded. Suppose also that (7.7.2) has an exponential dichotomy with data* $(\alpha, K_{ed}, P(\cdot))$ *and that condition* $(\mathcal{E}^1_{dif})$ *holds with* $g(\cdot, 0) \in l^1_\alpha(\mathbb{Z}^+; \mathbb{C}^r)$. *If*

$$K_{\mathcal{P}_{ps}} K_{ed} e^\alpha \sup_{m \geq 0} (1 + \|P(m)\|_{\mathcal{P}_{ps} \to \mathcal{P}_{ps}}) \|l_g\|_1 < 1, \tag{7.8.16}$$

then for any solution $z(n)$ *of (7.7.2) such that* $z_\bullet \in l^\infty_\alpha$, *there is a unique solution* $y(n)$ *of (7.8.3) such that* $y_\bullet \in l^\infty_\alpha$ *and*

$$y(n) = z(n) + o(e^{\alpha n}), \quad n \to \infty. \tag{7.8.17}$$

The converse is also true. Furthermore, the one-to-one correspondences $y_\bullet \mapsto z_\bullet$ *and* $z_\bullet \mapsto y_\bullet$ *are continuous.*

Proof. Let z be any solution of (7.7.2) such that $z_\bullet \in l^\infty_\alpha$. We define the operator $\Theta^\#$ on l^∞_α by

$$[\Theta^\# \eta](n) = z_n + \sum_{s=0}^\infty \Gamma(n, s) E^0(g(s, \eta(s))), \tag{7.8.18}$$

where $\eta \in l^\infty_\alpha$, $n \in \mathbb{Z}^+$, and $\Gamma(n, s)$ denotes the Green function associated with (7.7.2) [see (7.7.9)]. We have that

$$\left\| \sum_{s=0}^\infty \Gamma(n, s) E^0(g(s, \eta(s))) \right\|_{\mathcal{P}_{ps}}$$

$$\leq K_{\mathcal{P}_{ps}} K_{ed} \sup_{m \in \mathbb{Z}^+} (1 + \|P(m)\|_{\mathcal{P}_{ps} \to \mathcal{P}_{ps}})$$

$$\times \left\{ e^\alpha \sum_{s=0}^{n-1} e^{-\alpha n} e^{\alpha s} |g(s, \eta(s))| + e^{-\alpha} \sum_{s=n}^\infty e^{\alpha n} e^{-\alpha s} |g(s, \eta(s))| \right\}$$

$$\leq e^\alpha e^{\alpha n} K_{\mathcal{P}_{ps}} K_{ed} \sup_{m \in \mathbb{Z}^+} (1 + \|P(m)\|_{\mathcal{P}_{ps} \to \mathcal{P}_{ps}})$$

$$\times \left[\left(\sum_{s=0}^\infty l_g(s) \right) \|\eta\|_\alpha + \sum_{s=0}^\infty |g(s, 0)| e^{-\alpha s} \right].$$

Therefore

$$||\Theta^{\#}\eta||_{\alpha} \leq ||z_{\bullet}||_{\alpha}$$
$$+ K_{\mathcal{P}_{ps}} K_{ed} \sup_{m \in \mathbb{Z}^{+}} (1 + ||P(m)||_{\mathcal{P}_{ps} \to \mathcal{P}_{ps}}) \times (||l_g||_1 ||\eta||_{\alpha} + ||g(\cdot, 0)||_{1,\alpha}).$$

It proves that the space l_{α}^{∞} is invariant under $\Theta^{\#}$.

A similar argument shows that

$$||\Theta^{\#}\eta - \Theta^{\#}\xi|| \leq K_{\mathcal{P}_{ps}} K_{ed} e^{\alpha} \sup_{m \geq 0} (1 + ||P(m)||_{\mathcal{P}_{ps} \to \mathcal{P}_{ps}}) ||l_g||_1 ||\eta - \xi||_{\alpha}.$$

By (7.8.16) we find that $\Theta^{\#}$ is a contraction in l_{α}^{∞} and hence $\Theta^{\#}$ has a unique fixed point $\xi \in l_{\alpha}^{\infty}$.

Next we define

$$y(n) = \begin{cases} [\xi(n)](0) & n \geq 0, \\ [\xi(0)](n), & n < 0. \end{cases}$$

Applying [58, Lemma 2.8] we find that y is a solution of (7.8.3) and $y_n = \xi(n)$, $n \in \mathbb{Z}^{+}$. Conversely, let y be a solution of (7.8.3) such that $y_{\bullet} \in l_{\alpha}^{\infty}$. From [148, Theorem 2.1] we can verify that

$$y_n = T(n, 0)y_0 + [\Theta^{\#}y_{\bullet}](n), \quad n \in \mathbb{Z}^{+}.$$

Define a $\mathcal{P}_{ps}$-valued function ζ by

$$\zeta(n) = y_n - [\Theta^{\#}y_{\bullet}](n), \quad n \in \mathbb{Z}^{+}.$$

Whence $T(n, 0)\zeta(0) = \zeta(n), n \in \mathbb{Z}^{+}$.

Set

$$z(n) = \begin{cases} [\zeta(n)](0) & n \geq 0, \\ [\zeta(0)](n), & n < 0. \end{cases}$$

Applying [58, Lemma 2.8] we find that z is a solution of (7.7.2) and $z_n = \zeta(n)$, $n \in \mathbb{Z}^{+}$.

Next we prove the asymptotic relation (7.8.17). For $n \geq m$ we have

$$y_n = z_n + \sum_{s=0}^{m-1} T(n, s+1)P(s+1)E^0(g(s, y_s))$$

$$+ \sum_{s=m}^{\infty} \Gamma(n, s)E^0(g(s, y_s)). \tag{7.8.19}$$

We can assert that for m sufficiently large

$$\sum_{s=m}^{\infty} \Gamma(n,s) E^0(g(s,y_s)) = o(e^{\alpha n}), \quad n \to \infty. \tag{7.8.20}$$

In fact, we observe that

$$\left\| \sum_{s=m}^{\infty} \Gamma(n,s) E^0(g(s,y_s)) \right\|_{\mathcal{P}_{ps}} e^{-\alpha n}$$

$$\leq K_{\mathcal{P}_{ps}} K_{ed} e^{\alpha} \sup_{m \in \mathbb{Z}^+} (1 + \|P(m)\|_{\mathcal{P}_{ps} \to \mathcal{P}_{ps}})$$

$$\times \left\{ \left(\sum_{s=m}^{\infty} l_g(s) \right) \|y_\bullet\|_\alpha + \sum_{s=m}^{\infty} |g(s,0)| e^{-\alpha s} \right\}. \tag{7.8.21}$$

On the other hand, we can infer that

$$\left\| \sum_{s=0}^{m-1} T(n,s+1) P(s+1) E^0(g(s,y_s)) \right\|_{\mathcal{P}_{ps}} e^{-\alpha n}$$

$$\leq e^{\alpha(1+2m)} \left\{ \left(\sum_{s=0}^{m-1} l_g(s) \right) \|y_\bullet\|_\alpha + \sum_{s=0}^{m-1} |g(s,0)| e^{-\alpha s} \right\}.$$

$$\tag{7.8.22}$$

It is clear from expressions (7.8.19), (7.8.20), and (7.8.22) that we get (7.8.17). Finally we observe that

$$(1 - K_{\mathcal{P}_{ps}} K_T e^{\alpha} \sup_{m \geq 0} (1 + \|P(m)\|_{\mathcal{P}_{ps} \to \mathcal{P}_{ps}}) \|l_g\|_1) \|y_\bullet - \tilde{y}_\bullet\|_\alpha \leq \|z_\bullet - \tilde{z}_\bullet\|_\alpha$$

$$\leq (1 + K_{\mathcal{P}_{ps}} K_T e^{\alpha} \sup_{m \geq 0} (1 + \|P(m)\|_{\mathcal{P}_{ps} \to \mathcal{P}_{ps}}) \|l_g\|_1) \|y_\bullet - \tilde{y}_\bullet\|_\alpha.$$

This establishes the continuity of $y_\bullet \mapsto z_\bullet$ and $z_\bullet \mapsto y_\bullet$. The proof is therefore complete. $\qquad \square$

Now we state a local version of the Theorem 7.8.4.

Theorem 7.8.5 ([3]). *Assume that condition $(\mathcal{A}_{dif})$ is fulfilled and that the functions $N(\cdot)$ and $M(\cdot)$ given by Axiom (PS_1) are bounded. Furthermore, suppose that (7.7.2) has an exponential dichotomy with data $(\alpha, K_{ed}, P(\cdot))$ and assume that condition $(\mathcal{H}_{dif})$ holds with α instead of $\alpha^\#$. Then there are positive constants $\epsilon \in \mathbb{R}$ and $m \in \mathbb{Z}^+$ so that for any solution $z(n)$ of (7.7.2) for $n \geq m$ such* *that $z_{\bullet} \in \mathcal{L}_{\alpha}^{\infty}[\epsilon a]$, there exists a unique solution $y(n)$ of (7.8.3) for $n \geq m$ such that $y_{\bullet} \in \mathcal{L}_{\alpha}^{\infty}[a]$ and the asymptotic relation (7.8.17) holds. The converse is also true. Furthermore, the one-to-one correspondences $y_{\bullet} \mapsto z_{\bullet}$ and $z_{\bullet} \mapsto y_{\bullet}$ are continuous.*

7.9 Volterra Difference System with Infinite Delay

We apply our previous result to Volterra difference systems with infinite delay. Volterra difference equations can be considered as natural generalization of difference equations. During the last few years Volterra difference equations have emerged vigorously in several applied fields and nowadays there is a wide interest in developing the qualitative theory for such equations.[18]

Let $A(n)$, $K(m)$ be $r \times r$ matrices defined for $n, m \in \mathbb{Z}^{+}$, and let $\beta : \mathbb{Z}^{+} \longrightarrow \mathbb{R}^{+}$ be an arbitrary positive increasing sequence such that

$$\sum_{n=0}^{\infty} |K(n)|\beta(n) < +\infty. \tag{7.9.1}$$

We consider the following Volterra difference system with infinite delay:

$$x(n+1) = \sum_{s=-\infty}^{n} A(n)K(n-s)x(s), \quad n \geq 0. \tag{7.9.2}$$

This equation is viewed as a functional difference equation on the phase space $\mathcal{P}_{ps}^{\beta,r}$, where $\mathcal{P}_{ps}^{\beta,r}$ is defined as follows:

$$\mathcal{P}_{ps}^{\beta,r} = \mathcal{P}_{ps}^{\beta,r}(\mathbb{Z}^{-};\mathbb{C}^{r}) = \{\varphi : \mathbb{Z}^{-} \longrightarrow \mathbb{C}^{r} : \mathrm{Sup}_{n \in \mathbb{Z}^{+}} |\varphi(-n)|/\beta(n) < +\infty\}, \tag{7.9.3}$$

with the norm

$$\|\varphi\|_{\mathcal{P}_{ps}^{\beta,r}} = \mathrm{Sup}_{n \in \mathbb{Z}^{+}} |\varphi(-n)|/\beta(n), \varphi \in \mathcal{P}_{ps}^{\beta,r}. \tag{7.9.4}$$

We have the following result as a consequence of Theorem 7.7.6.

Theorem 7.9.1 ([59]). *Assume that system (7.9.2) has an exponential dichotomy on $\mathbb{N}$ with data $(\alpha, K_{ed}, P(n))$. Then, for any $h \in \ell^{p}(\mathbb{N})$ (with $1 \leq p \leq +\infty$) and any $\varphi \in \mathrm{Range}(P(0))$, the boundary value problem*

---

[18]See [1, 3, 15, 40, 41, 50, 56, 60, 92, 93, 122, 123, 125, 172, 173] and references therein.

$$z(n+1) = \sum_{s=-\infty}^{n} A(n)K(n-s)z(s) + h(n), \qquad (7.9.5)$$

$$P(0)z_0 = \varphi, \qquad (7.9.6)$$

has a unique solution $z_\bullet \in l^p(\mathbb{Z}^+; \mathcal{P}_{ps}^{\beta,r})$, namely $z = z^{sp} + z^{hom}$, where

$$z_n^{sp} = \sum_{s=0}^{\infty} \Gamma(n,s)E^0(h(s)), \qquad (7.9.7)$$

$$z_n^{hom} = T(n,0)P(0)\varphi, \qquad (7.9.8)$$

and $\Gamma(n,s)$ is the Green function associated with (7.9.2). On the other hand, the solution z satisfies $z \in \ell^{p'}(\mathbb{N})$ for all $1 \leq p \leq p' \leq +\infty$ and the following estimates hold:

$$(1-e^{-\alpha})^{1-\frac{1}{p}+\frac{1}{p'}} ||z_\bullet^{sp}||_{p'} + (1-e^{-\alpha})^{1-\frac{1}{p}} ||z_0^{sp}||_{\mathcal{P}_{ps}^{\beta,r}} \leq 4KK_{ed}||h||_p, \quad (7.9.9)$$

$$(1-e^{-\alpha})^{\frac{1}{p'}} ||z_\bullet^{hom}||_{p'} + ||z_0^{hom}||_{\mathcal{P}_{ps}^{\beta,r}} \leq 2K_{ed}||\varphi||_{\mathcal{P}_{ps}^{\beta,r}}, \qquad (7.9.10)$$

where α and K_{ed} are the constants of Definition 7.7.1.

Remark 7.9.2. Note that in the preceding estimates, we get $\frac{1}{p} = 0$ for $p = +\infty$.

Example 7.9.3. Let $a_i(n)$, $i = 1,2$ be two sequences and σ, α, γ be three positive constants such that[19]:

(i) $\rho_1^* := \mathrm{Sup}_{n \geq 0} \max_{-n \leq \theta \leq 0} \left[\left[\prod_{s=n+\theta}^{n-1} |a_1(s)|^{-1} \right] / e^{-\gamma\theta} \right] < +\infty.$

(ii) $\prod_{s=\tau}^{n-1} |a_1(s)| \leq \sigma e^{-\alpha(n-\tau)}, \quad n \geq \tau \geq 0.$

(iii) $\prod_{s=n}^{\tau-1} |a_2(s)|^{-1} \leq \sigma e^{-\alpha(\tau-n)}, \quad \tau \geq n \geq 0.$

---

[19]Some concrete examples of functions a_1 and a_2 satisfying the previous assumptions are:

(a) $a_1(n) := 1/\delta$, $a_2(n) := \delta$ with $1 < \delta \leq e^\gamma$ or $1/\mu \leq \delta \leq \nu e^\gamma$, where $\mu, \nu \in (0,1)$.
(b) $\eta < e^{-\gamma} < \mu < 1, \gamma > 0$, $a_1(n) := \mu$, $a_2(n) := 1/\eta$.
(c) $1/\nu e^\gamma \leq |a_1(n)| \leq \mu, 1/\mu \leq |a_2(n)|$, for all $n \geq 0$, where $\mu, \nu \in (0,1)$.

From now until the end of Example 7.9.3, we will assume that a_1 and a_2 are functions satisfying (i)–(iii). Using (ii) and (iii), we can assert that

$$\prod_{s=\tau}^{n-1} |a_2(s)|^{-1} \leq \sigma^2 \prod_{s=\tau}^{n-1} |a_1(s)|^{-1}, \quad n \geq \tau. \tag{7.9.11}$$

We consider the following nonautonomous difference system:

$$x(n+1) = A(n)x(n), \tag{7.9.12}$$

where $A(n)$ is a 2×2 matrix defined by $diag(a_1(n), a_2(n))$. For convenience of the reader, we would like to begin with a complete analysis to check the dichotomic properties.

We recall that the solution operator $T(n, \tau)$, $n \geq \tau$, of (7.9.12) is a bounded linear operator on the phase space $\mathcal{P}_{ps}^{\beta,2} = \mathcal{P}_{ps}^{\beta,2}(\mathbb{Z}^-; \mathbb{C}^2)$, with $\beta(n) = e^{\gamma n}$, and is defined by

$$[T(n, \tau)\varphi](\theta)$$
$$= \begin{cases} \left(\left(\prod_{s=\tau}^{n+\theta-1} a_1(s)\right)\varphi^1(0), \left(\prod_{s=\tau}^{n+\theta-1} a_2(s)\right)\varphi^2(0)\right), -(n-\tau) \leq \theta \leq 0, \\ (\varphi^1(n-\tau+\theta), \varphi^2(n-\tau+\theta)), \quad \theta \leq -(n-\tau). \end{cases}$$

A computation shows that

$$T(n, s)T(s, m) = T(n, m), \; n \geq s \geq m \quad \text{and} \quad T(m, m) = I.$$

We define the projections $P(n), Q(n) : \mathcal{P}_{ps}^{\beta,2} \longrightarrow \mathcal{P}_{ps}^{\beta,2}$ by

$$[P(n)\varphi](\theta) = \begin{cases} \left(\varphi^1(\theta), \varphi^2(\theta) - \left(\prod_{s=n+\theta}^{n-1} a_2(s)^{-1}\right)\varphi^2(0)\right), \quad -n \leq \theta \leq 0, \\ (\varphi^1(\theta), \varphi^2(\theta)), \quad \theta < -n, \end{cases}$$

and $Q(n) = I - P(n) : \mathcal{P}_{ps}^{\beta,2} \longrightarrow \mathcal{P}_{ps}^{\beta,2}$.

For $n \geq \tau$ we observe that $T(n, \tau) : Q(\tau)\mathcal{P}_{ps}^{\beta,2} \longrightarrow Q(n)\mathcal{P}_{ps}^{\beta,2}$ is given by

$$[T(n, \tau)Q(\tau)\varphi](\theta) = \begin{cases} \left(0, \left(\prod_{s=\tau}^{n+\theta-1} a_2(s)\right)\varphi^2(0)\right), \quad -(n-\tau) \leq \theta \leq 0, \\ \left(0, \left(\prod_{s=n+\theta}^{\tau-1} a_2(s)^{-1}\right)\varphi^2(0)\right), \quad -n \leq \theta \leq -(n-\tau), \\ (0, 0), \quad \theta < -n. \end{cases}$$

We find that for $n \geq \tau$

$$T(n, \tau)Q(\tau) = Q(n)T(n, \tau), \quad T(n, \tau)P(\tau) = P(n)T(n, \tau).$$

We can prove that $T(n, \tau)$, $n \geq \tau$, is an isomorphism from $Q(\tau)\mathcal{P}_{ps}^{\beta,2}$ onto $Q(n)\mathcal{P}_{ps}^{\beta,2}$. We define $T(\tau, n)$ as the inverse mapping, which is given by

$$[T(\tau, n)Q(n)\varphi](\theta) = \begin{cases} \left(0, \left(\displaystyle\prod_{s=\tau+\theta}^{n-1} a_2(s)^{-1}\right)\varphi^2(0)\right), & -\tau \leq \theta \leq 0, \\ (0, 0), & \theta < -\tau. \end{cases}$$

By virtue of (7.9.11), we claim that there is positive constant K_{ed} such that

$$\|T(n, \tau)P(\tau)\|_{\mathcal{P}_{ps}^{\beta,2} \to \mathcal{P}_{ps}^{\beta,2}} \leq K_{ed}e^{-\alpha(n-\tau)}, \quad n \geq \tau. \tag{7.9.13}$$

In fact,

$$\|T(n, \tau)P(\tau)\|_{\mathcal{P}_{ps}^{\beta,2} \to \mathcal{P}_{ps}^{\beta,2}}$$

$$\leq \max_{-(n-\tau)\leq\theta\leq 0}\left[\left[\prod_{s=\tau}^{n+\theta-1}|a_1(s)|\right]e^{\gamma\theta}\right]$$

$$+3\max_{-n\leq\theta\leq-(n-\tau)}\left[\left[\prod_{s=n+\theta}^{\tau-1}|a_2(s)|^{-1}\right]e^{\gamma\theta}\right]$$

$$\leq \left[\prod_{s=\tau}^{n-1}|a_1(s)|\right]\max_{-(n-\tau)\leq\theta\leq 0}\left[\left[\prod_{s=n+\theta}^{n-1}|a_1(s)|^{-1}\right]e^{\gamma\theta}\right]$$

$$+3\sigma^2\left[\prod_{s=\tau}^{n-1}|a_1(s)|\right]\max_{-n\leq\theta\leq-(n-\tau)}\left[\left[\prod_{s=n+\theta}^{n-1}|a_1(s)|^{-1}\right]e^{\gamma\theta}\right]$$

$$\leq 4\sigma^2\rho_1^*\left[\prod_{s=\tau}^{n-1}|a_1(s)|\right].$$

On the other hand, we can verify that

$$\|T(n, \tau)Q(\tau)\|_{\mathcal{P}_{ps}^{\beta,2} \to \mathcal{P}_{ps}^{\beta,2}} \leq \sigma\rho_2^*e^{-\alpha(\tau-n)}, \quad \tau \geq n, \tag{7.9.14}$$

where

$$\rho_2^* := \mathrm{Sup}_{n\geq 0} \max_{-n\leq\theta\leq 0}\left[\Big[\prod_{s=n+\theta}^{n-1}|a_2(s)|^{-1}\Big]\Big/e^{-\gamma\theta}\right].$$

Therefore system (7.9.12) has an exponential dichotomy.

For any $h \in \ell^p(\mathbb{Z}^+;\mathbb{C}^2)$ (with $1 \leq p \leq +\infty$) and any $\varphi \in Range(P(0))$, Theorem 7.9.1 assures that the boundary value problem

$$z(n+1) = A(n)z(n) + h(n), \quad n \geq 0, \tag{7.9.15}$$

$$P(0)z_0 = \varphi, \tag{7.9.16}$$

has a unique solution z so that $z_\bullet \in l^p(\mathbb{Z}^+;\mathcal{P}_{ps}^{\beta,2})$. Moreover $z \in \ell^{p'}(\mathbb{Z}^+;\mathbb{C}^2)$ for all $1 \leq p \leq p' \leq +\infty$ and the estimates (7.9.9) and (7.9.10) hold.

We note that the projectors $P(n)$ are not unique, but the ranges are unique (see Proposition 7.7.4 for more details). It is worth noting that one can construct other projectors $\hat{P}(n)$ from $P(n)$ such that (7.9.12) has an exponential dichotomy. Following the general method established in Proposition 7.7.4 we construct new projectors

$$[\hat{P}(n)\varphi](\theta) =$$
$$\begin{cases} \left(\left(\varphi^1(\theta) + \Big(\prod_{s=0}^{n+\theta-1} a_1(s)\Big)\Big(\prod_{s=0}^{n-1} a_2(s)^{-1}\Big)\varphi^2(0), \varphi^2(\theta)\right.\right. \\ \quad\left.\left. - \Big(\prod_{s=n+\theta}^{n-1} a_2(s)^{-1}\Big)\varphi^2(0)\right)\right) & \text{if } -n \leq \theta \leq 0, \\ (\varphi^1(\theta), \varphi^2(\theta)), & \text{if } \theta < -n, \end{cases}$$

such that (7.9.12) has an exponential dichotomy.

Putting $\hat{Q}(n) = I - \hat{P}(n) : \mathcal{P}_{ps}^{\beta,2} \longrightarrow \mathcal{P}_{ps}^{\beta,2}$, we can prove that $T(n,\tau), n \geq \tau$, is an isomorphism from $\hat{Q}(\tau)\mathcal{P}_{ps}^{\beta,2}$ onto $\hat{Q}(n)\mathcal{P}_{ps}^{\beta,2}$. We define $T(\tau,n)$ as the inverse mapping, which is given by

$$[T(\tau,n)\hat{Q}(n)\varphi](\theta) =$$
$$\begin{cases} \left(\left(-\Big(\prod_{s=0}^{\tau+\theta-1} a_1(s)\Big)\Big(\prod_{s=\tau}^{n-1} a_2(s)^{-1}\Big)\varphi^2(0), \Big(\prod_{s=\tau+\theta}^{n-1} a_2(s)^{-1}\Big)\varphi^2(0)\right)\right), & -\tau \leq \theta \leq 0, \\ \\ (0,0), & \theta < -\tau. \end{cases}$$

We have the following estimates:

$$\|T(n,\tau)\hat{P}(\tau)\|_{\mathcal{P}_{ps}^{\beta,2}\to\mathcal{P}_{ps}^{\beta,2}}$$

$$\leq (1+\sigma^2)\max_{-(n-\tau)\leq\theta\leq0}\left[\left[\prod_{s=\tau}^{n+\theta-1}|a_1(s)|\right]e^{\gamma\theta}\right]$$

$$+2e^{\gamma(\tau-n)}+\max_{-n\leq\theta\leq-(n-\tau)}\left[\left[\prod_{s=n+\theta}^{\tau-1}|a_2(s)|^{-1}\right]e^{\gamma\theta}\right]$$

$$+\max_{-n\leq\theta\leq-(n-\tau)}\left[\left[\prod_{s=0}^{n+\theta-1}|a_1(s)|\right]\left[\prod_{s=0}^{\tau-1}|a_2(s)|^{-1}\right]e^{\gamma\theta}\right]$$

$$\leq (1+\sigma^2)\max_{-(n-\tau)\leq\theta\leq0}\left[\left[\prod_{s=\tau}^{n+\theta-1}|a_1(s)|\right]e^{\gamma\theta}\right]$$

$$+3\max_{-n\leq\theta\leq-(n-\tau)}\left[\left[\prod_{s=n+\theta}^{\tau-1}|a_2(s)|^{-1}\right]e^{\gamma\theta}\right]$$

$$+\max_{-n\leq\theta\leq-(n-\tau)}\left[\left[\prod_{s=0}^{n+\theta-1}|a_1(s)|\right]\left[\prod_{s=0}^{\tau-1}|a_2(s)|^{-1}\right]e^{\gamma\theta}\right]$$

$$\leq (1+\sigma^2)\left[\prod_{s=\tau}^{n-1}|a_1(s)|\right]\max_{-(n-\tau)\leq\theta\leq0}\left[\left[\prod_{s=n+\theta}^{n-1}|a_1(s)|^{-1}\right]e^{\gamma\theta}\right]$$

$$+4\sigma^2\left[\prod_{s=\tau}^{n-1}|a_1(s)|\right]\max_{-n\leq\theta\leq-(n-\tau)}\left[\left[\prod_{s=n+\theta}^{n-1}|a_1(s)|^{-1}\right]e^{\gamma\theta}\right]$$

$$\leq (1+5\sigma^2)\rho_1^*\prod_{s=\tau}^{n-1}|a_1(s)|.$$

On the other hand, we can verify that

$$\|T(n,\tau)\hat{Q}(\tau)\|_{\mathcal{P}_{ps}^{\beta,2}\to\mathcal{P}_{ps}^{\beta,2}}\leq(\sigma\rho_1^*+\rho_2^*)\sigma e^{-\alpha(\tau-n)},\quad \tau\geq n,\qquad (7.9.17)$$

From the last two estimates, we find that (7.9.12) has an exponential dichotomy. This finishes the discussion of Example 7.9.3.

Let γ be a positive real number and let $A(n)$, $K(n)$ and $D(n,s)$ be three $r\times r$ matrices defined for $n\in\mathbb{Z}^+$, $s\in\mathbb{Z}$ such that (7.9.1) holds with $\beta(n)=e^{\gamma n}$ and

$$\|A\|_\infty=\sup_{n\geq0}|A(n)|<\infty.\qquad (7.9.18)$$

We consider the following Volterra difference system with infinite delay:

$$y(n+1) = \sum_{s=-\infty}^{n} [A(n)K(n-s) + vD(n,s)]y(s), \quad n \geq 0, \qquad (7.9.19)$$

where v is a real number.

We recall that the Volterra system (7.9.19) is viewed as a retarded functional difference equation on the phase space $\mathcal{P}_{ps}^{\beta,r}$.

We have, as a consequence of Theorem 7.7.7, the following result.

Theorem 7.9.4 ([3]). *Suppose that the following hypothesis hold:*

1. *System (7.9.2) possesses an exponential dichotomy.*
2. *There is a sequence $\beta^{\natural} \in l^p(\mathbb{Z}^+)$ such that*

$$\sum_{\tau=0}^{n} |D(n,\tau)| + \sum_{\tau=-\infty}^{-1} |D(n,\tau)|e^{-\gamma\tau} \leq \beta^{\natural}(n), \quad n \geq 0.$$

If $|v|$ is small enough, then for each $\varphi \in P(0)\mathcal{P}_{ps}^{\beta,r}$ there is a unique bounded solution y of the system (7.9.19) with $P(0)y_0 = \varphi$ such that $y_{\bullet} \in l^p(\mathbb{Z}^+;\mathcal{P}_{ps}^{\beta,r})$, in particular $y \in l^p(\mathbb{Z}^+;\mathbb{C}^r)$. Moreover, we have the following a priori estimate for the solution:

$$\|y_{\bullet}\|_p \leq C\|\varphi\|_{\mathcal{P}_{ps}^{\beta,r}},$$

where $C > 0$ is a suitable constant. Furthermore, the application $\varphi \in P(0)\mathcal{P}_{ps}^{\beta,r} \mapsto y_{\bullet}(\varphi) \in l^p(\mathbb{Z}^+;\mathcal{P}_{ps}^{\beta,r})$ is continuous.

Let $B(n)$ and $G(s)$ be two $r \times r$ matrices defined for $n \in \mathbb{Z}^+$ and $s \in \mathbb{Z}^-$ such that $|B(\cdot)| \in l^p$ and

$$\sum_{n=0}^{\infty} |G(-n)|\beta(n) < \infty. \qquad (7.9.20)$$

Next we consider the following Volterra difference system with infinite delay:

$$y(n+1) = \sum_{s=-\infty}^{n} [A(n)K(n-s) + B(n)G(s-n)|y(0)|]y(s), \quad n \geq 0. \quad (7.9.21)$$

As a consequence of Theorem 7.7.9 we have the following result.

Theorem 7.9.5. *Suppose that system (7.9.2) has an exponential dichotomy. Then there are positive constants $\tilde{M} \in \mathbb{R}$ and $m \in \mathbb{Z}^+$ such that for each $\varphi \in$*

$P(m)\mathcal{P}_{ps}^{\beta,r}$ with $\|\varphi\|_{\mathcal{P}_{ps}^{\beta,r}} \leq \tilde{M}$, there is a unique bounded solution y of the Volterra system (7.9.21) for $n \geq m$ with $P(m)y_m = \varphi$ such that $y_n = o(1)$ as $n \to \infty$.

Example 7.9.6. We shall use the same notations as in Example 7.9.3. We consider the following perturbation of (7.9.12):

$$x(n+1) = L_1(n, x_n), \quad n \geq 0, \tag{7.9.22}$$

where

$$L_1(n, \varphi) = A(n)\varphi(0) + B(n)\varphi(-1), \quad \varphi \in \mathcal{P}_{ps}^{\beta,2},$$

and $B(n)$ is a 2×2 matrix with $\|B(\cdot)\|_\infty$ sufficiently small. Then by the Cardoso–Cuevas' perturbation theorem [36, Theorem 1.3] (see also comments in Sect. 7.10, Theorem 7.10.2), equation (7.9.22) has an exponential dichotomy for suitable data $(\tilde{\alpha}, \tilde{K}_{ed}, \tilde{P}(\cdot))$.

Let $D(n)$ be a 2×2 matrix defined for $n \in \mathbb{Z}^+$ such that $\sum_{n=0}^{\infty} |D(n)|^p < \infty$. We have, as a consequence of Theorem 7.9.4, the following result.

Proposition 7.9.7. *Let v be a real number such that $|v|$ is small enough, and $\varphi \in \tilde{P}(0)\mathcal{P}_{ps}^{\beta,2}$. Then*

$$x(n+1) = L_1(n, x_n) + vD(n)x(n), \quad n \geq 0, \tag{7.9.23}$$

has a unique bounded solution $y(n)$ with $\tilde{P}(0)y_0 = \varphi$ such that $y \in l^p(\mathbb{Z}^+, \mathbb{C}^2)$.

Next we consider the equation

$$x(n+1) = L_1(n, x_n) + D(n)|x(0)|x(n), \quad n \geq 0. \tag{7.9.24}$$

By Theorem 7.7.9, there are positive constants $M^\odot \in \mathbb{R}$ and $m \in \mathbb{Z}^+$ such that for each $\varphi \in \tilde{P}(m)\mathcal{P}_{ps}^{\beta,2}$ with $\|\varphi\|_{\mathcal{P}_{ps}^{\beta,2}} \leq M^\odot$, there is a unique bounded solution y of (7.9.24) for $n \geq m$ with $\tilde{P}(m)y_m = \varphi$ such that the map $n \mapsto y_n$ belongs to $l^p(\mathbb{Z}^+; \mathcal{P}_{ps}^{\beta,2})$. This finishes the discussion of Example 7.9.6.

7.10 Comments

For fundamental theory of stability and applications, we refer the reader to, for example, the works of Elaydi [75], Gajić and Qureshi [85], Hahn and Parks [95], Kelly and Peterson [116], and Kocic and Ladas [121].

Theorem 7.1.1 is proved in [39]. Theorem 7.3.1 is taken from [37]. Theorem 7.4.1 (resp. Theorem 7.4.8) and Corollary 7.4.2 (resp. Corollaries 7.4.9 and 7.4.10) are from [54] (resp. [37]). We emphasize here that it is not hard to establish a local version of Theorem 7.6.1.

For the basic theory of phase spaces, the reader is referred to the book by Hino et al. [99]. The abstract phase spaces were introduced by Hale and Kato [96] for studying qualitative theory of functional differential equations with unbounded delay. The idea of considering phase spaces for studying qualitative properties of functional difference equations was used first by Murakami [149]. He also used phase spaces to study some spectral properties of the solution operator for linear Volterra difference systems. Phase spaces were later used by Elaydi et al. [76] to study asymptotic equivalence of bounded solutions of a homogeneous Volterra difference system and its perturbation.

Volterra difference equations mainly arise in modeling many real-world phenomena, for example, in the study of competitive species in population dynamics and the study of motions of interacting bodies and on applying numerical methods for solving Volterra integral or integrodifferential equations. It is to be noted that Volterra systems describe a process whose current state is determined by their entire prehistory. These processes are encountered, for example, in models of propagation of perturbation in materials with memory and various models to describe the evolution of epidemics, the theory of viscoelasticity, and the study of optimal control problems (see [75, 122–125, 136], and the references therein).

Exponential Dichotomy

Let $x(\cdot, 0, \varphi)$ be the solution of the homogeneous linear system (7.7.2) passing through $(0, \varphi)$; $x.(0, \varphi)$ denotes the $\mathcal{P}_{ps}^{\gamma}$-valued function defined by $n \longrightarrow x_n(0, \varphi)$ (see (7.7.4) for the definition of $\mathcal{P}_{ps}^{\gamma}$). We can infer that $x.(0, \varphi)$ is the solution of the following equation:

$$Z(n + 1) = T(n + 1, n)Z(n), \quad n \geq 0, \tag{7.10.1}$$

where $T(n, \tau) : \mathcal{P}_{ps}^{\gamma} \rightarrow \mathcal{P}_{ps}^{\gamma}, n \geq \tau$, denotes the solution operator of the homogeneous linear system (7.7.2). For any numbers $1 \leq p < +\infty$, we define the l_p-stable space $\mathcal{P}_{ps}^{\circ}(\tau)$, $\tau \in \mathbb{Z}^+$, by

$$\mathcal{P}_{ps}^{\circ}(\tau) := \left\{ \varphi \in \mathcal{P}_{ps}^{\gamma} : \sum_{n=\tau}^{\infty} ||T(n, \tau)\varphi||_{\mathcal{P}_{ps}^{\gamma}}^{p} < +\infty \right\}.$$

For $p = +\infty$, we define $\mathcal{P}_{ps}^{\circ}(\tau) := \left\{ \varphi \in \mathcal{P}_{ps}^{\gamma} : \sup_{n \geq \tau} ||T(n, \tau)\varphi||_{\mathcal{P}_{ps}^{\gamma}} < +\infty \right\}$. An orbit $T(n, \tau)\varphi$, for $n \geq \tau \geq 0$ and $\varphi \in \mathcal{P}_{ps}^{\circ}(\tau)$, is called a l_p-stable orbit.

In order to find conditions for (7.7.2) to have an exponential dichotomy we relate this property to the solvability of the following inhomogeneous equation:

$$Z(n + 1) = T(n + 1, n)Z(n) + f(n), \quad n \geq 0, \tag{7.10.2}$$

in some spaces for each f. In other words, one wants to relate the exponential dichotomy of (7.7.2) to the surjectiveness of the operator Υ defined by

$$(\Upsilon\xi)(n) = \xi(n+1) - T(n+1,n)\xi(n), \qquad (7.10.3)$$

for ξ belonging to a suitable space. More precisely, we have the following result of Perron type due to Cardoso and Cuevas [36].

Theorem 7.10.1. *Assume that* $(\mathcal{A}_{dif})$ *is fulfilled in the space* $\mathcal{P}_{ps}^{\gamma}$. *Then for every* $1 \leq p \leq +\infty$, *the following assertions are equivalent:*

(i) *The equation* (7.7.2) *has an exponential dichotomy on* $\mathcal{P}_{ps}^{\gamma}$.
(ii) $\Upsilon : l^p \longrightarrow l^p$ *is surjective and* $\mathcal{P}_{ps}^{\circ}(0)$ *is complemented in* $\mathcal{P}_{ps}^{\gamma}$.

The Robustness of Exponential Dichotomy

We have the following perturbation theorem (see [36, Theorem 1.3]).

Theorem 7.10.2. *Assume that* $(\mathcal{A}_{dif})$ *is fulfilled in the space* $\mathcal{P}_{ps}^{\gamma}$ *and suppose that* (7.7.2) *has an exponential dichotomy on* $\mathcal{P}_{ps}^{\gamma}$. *Furthermore let* $\{\mathcal{I}(n,\cdot)\}_{n\in\mathbb{Z}^+}$ *be a sequence of bounded linear operators from* $\mathcal{P}_{ps}^{\gamma}$ *into* $\mathbb{C}^r$. *If* $\mathcal{H} :=$ $\sup_{n\in\mathbb{Z}^+} ||\mathcal{I}(n,\cdot)||_{\mathcal{P}_{ps}^{\gamma}\rightarrow\mathbb{C}^r}$ *is sufficiently small, then the equation*

$$x(n+1) = L(n,x_n) + \mathcal{I}(n,x_n), \quad n \geq 0, \qquad (7.10.4)$$

has an exponential dichotomy as well.

References

1. R.P. Agarwal, *Difference Equations and Inequalities*, Monographs and Textbooks in Pure and Applied Mathematics, vol. 228 (Marcel Dekker, New York, 2000)
2. R.P. Agarwal, A. Cabada, V. Otero-Espinar, S. Dontha, Existence and uniqueness of solutions for anti-periodic difference equations. Archiv. Inequal. Appl. **2**, 397–412 (2002)
3. R.P. Agarwal, C. Cuevas, M. Frasson, Semilinear functional difference equations with infinite delay. Math. Comput. Model. **55**, 1083–1105 (2012)
4. H. Amann, Operator-valued Fourier multipliers, vector-valued Besov spaces, and applications. Math. Nachr. **186**, 5–56 (1997)
5. H. Amann, Maximal regularity for nonautonomous evolution equations. Adv. Nonlinear Stud. **4**(4), 417–430 (2004)
6. H. Amann, in *Quasilinear Parabolic Functional Evolution Equations*, ed. by M. Chipot, H. Ninomiya, Recent Advances in Elliptic and Parabolic Issues. Proceeding of the 2004 Swiss-Japanese Seminar (World Scientific, Singapore, 2006) pp. 19–44
7. H. Amann, *Linear and Quasilinear Parabolic Problems*, Monographs in Mathematics, vol. 89, (Basel, Birkhäuser Verlag, 1995)
8. W. Arendt, S. Bu, The operator-valued Marcinkiewicz multiplier theorem and maximal regularity. Math. Z. **240**, 311–343 (2002)
9. W. Arendt, S. Bu, Operator-valued Fourier multiplier on periodic Besov spaces and applications. Proc. Edin. Math. Soc. **47**(2), 15–33 (2004)
10. W. Arendt, *Semigroups and Evolution Equations: Functional Calculus, Regularity and Kernel Estimates.* Evolutionary Equations, vol. 1, Handbook of Differential Equation (North-Holland, Amsterdam, 2004) pp. 1–85
11. W. Arendt, M. Duelli, Maximal L^p–regularity for parabolic and elliptic equations on the line. J. Evol. Equat. **6**, 773–790 (2006)
12. A. Ashyralyev, C. Cuevas, S. Piskarev, On well-posedness of difference schemes for abstract elliptic problems in $L^p([0, T]; E)$ spaces. Numer. Funct. Anal. Optim. **29**(1–2), 43–65 (2008)
13. A. Ashyralyev, S. Piskarev, L. Weis, On well-posedness of difference schemes for abstract parabolic equations in $L^p([0, T]; E)$ spaces. Numer. Funct. Anal. Optim. **23**(7–8), 669–693 (2002)
14. J.B. Baillon, Caractère borné de certains générateurs de semigroupes linéaires dans les espaces de Banach. C. R. Acad. Sci. Paris Série A **290**, 757–760 (1980)
15. C.T.H. Baker, Y. Song, Periodic solutions of discrete Volterra equations. Math. Comput. Simulat. **64**(5), 521–542 (2004)
16. J.R. Barry, E.A. Lee, D.G. Messerschmitt, *Digital Communication*, 3rd ed. (Kluwer Academic Publishers, Boston, MA 2003)
17. A. Bátkai, E. Fasanga, R. Shvidkoy, Hyperbolicity of delay equations via Fourier multiplier. Acta Sci. Math. (Szeged), **69**, 131–145 (2003)

18. J. Bergh, J. Löfström, *Interpolation Spaces. An Introduction*, Grundlehren der Mathematischen Weissenchaften, vol. 223 (Springer, Berlin, 1976)
19. E. Berkson, T.A. Gillespie,Spectral decompositions and harmonic analysis on UMD-spaces, Studia Math. **112**(1), 13–49 (1994)
20. O.V. Besov, On a certain family of functional spaces, embedding and continuation. Dokl. Akad. Nauk SSSR **126**, 1163–1165 (1956)
21. W.J. Beyn, J. Lorenz, Stability of traveling waves: Dichotomies and eingenvalue conditions on finite intervals. Numer. Funct. Anal. Optimiz. **20** 201–244 (1999)
22. S. Blunck, Maximal regularity of discrete and continuous time evolution equations. Studia Math. **146**(2), 157–176 (2001)
23. S. Blunck, Analyticity and discrete maximal regularity of L_p-spaces. J. Funct. Anal. **183**(1), 211–230 (2001)
24. J. Bourgain, Some remarks on Banach spaces in which martingale differences sequences are unconditional. Arkiv Math. **21**, 163–168 (1983)
25. J. Bourgain, *Vector-Valued Singular Integrals and the H^1-BMO Duality*, Probability Theory and Harmonic Analysis (Marcel Dekker, New York, 1986)
26. L. Brand, *Differential and Difference Equations* (John Wiley, New York, 1966)
27. S. Bu, Y. Fang, Maximal regularity of second order delay equations in Banach spaces. Sci. China Math. **53**(1), 51–62 (2010)
28. S. Bu, J. Kim, Operator-valued Fourier multiplier on periodic Triebel spaces. Acta Math. Sin. (Engl. Ser.) **21**, 1049–1056 (2004)
29. S. Bu, Maximal regularity of second order delay equations in Banach spaces. Acta Math. Sin. (Engl. Ser.) **25**, 21–28 (2009)
30. D.L. Burkhölder, A geometric characterization of Banach spaces in which martingale differences are unconditional. Ann. Probab. **9**, 997–1011 (1981)
31. D.L. Burkhölder, *Martingale Transforms and the Geometry of Banach Spaces*, Lecture Notes in Mathematics, vol. 860, (Springer, Berlin, 1981), p. 35–50
32. D.L. Burkhölder, *A Geometrical Condition that Implies the Existence of Certain Singular Integrals on Banach-Space-Valued Functions*, Conference on Harmonic Analysis in Honour of Antoni Zygmund, Chicago 1981, ed. by W. Becker, A.P. Calderón, R. Fefferman, P.W. Jones, (Wadsworth, Belmont 1983), pp. 270–286.
33. D.L. Burkhölder, *Exploration in Martingales and its Applications*, Lecture Notes in Mathematics, vol. 1464, (Springer, 1991), pp. 1–66
34. D.L. Burkhölder, *Martingales and Singular Integrals in Banach Spaces*, Handbook of the Geometry of Banach Spaces, vol. 1, ed. by W.B. Johnson, J. Lindenstrauss (Elsevier, Amsterdam, 2001)
35. J.A. Cadzow, *Discrete Time Systems* (Printice Hall, New Jersey, 1973)
36. F. Cardoso, C. Cuevas, Exponential dichotomy and boundedness for retarded functional difference equations. J. Differ. Equat. Appl. **15**(3), 261–290 (2009)
37. A. Castro, C. Cuevas, Perturbation theory, stability, boundedness and asymptotic behavior for second order evolution equations in discrete time. J. Differ. Equat. Appl. **17**, 327–358 (2011)
38. A. Castro, C. Cuevas, C. Lizama, Maximal regularity of the discrete harmonic oscillator equation. Adv. Differ. Equat. **2009**, 1–14 (2009), Article ID 290625
39. A. Castro, C. Cuevas, C. Lizama, Well-posedness of second order evolution equation on discrete time. J. Differ. Equat. Appl. **16**, 1165–1178 (2010)
40. A. Castro, C. Cuevas, F. Dantas, H. Soto, About the behavior of solutions for Volterra difference equations with infinite delay. J. Comput. Appl. Math. **255**, 44–59 (2013)
41. S. Choi, N. Koo, Asymptotic property of linear Volterra difference systems. J. Math. Anal. Appl. **321**(1), 260–272 (2006)
42. J.L. Cieśliński, B. Ratkiewicz, On simulations of the classical harmonic oscillator equation by difference equations. Adv. Difference Equat. **2006**, (2006) doi:10.1155/ADE/2006/40171.
43. J.L. Cieśliński, On the exact discretization of the classical harmonic oscillator equation. J. Differ. Equat. Appl. **17**(11), 1673–1694 (2011)

44. J.L. Cieśliński, B. Ratkiewicz, Long-time behavior of discretizations of the simple pendulum. J. Phy. A: Math. Theor. **42**(105204), 29 (2009) doi 10.1088/1751-8113/42/10/105204.
45. Ph. Clément, B. de Pagter, F.A. Sukochev, M. Witvliet, Schauder decomposition and multiplier theorems. Studia Math. **138**, 135–163 (2000)
46. Ph. Clément, S.O. Londen, G. Simonett, Quasilinear evolutionary equations and continuous interpolation spaces. J. Differ. Equat. **196**(2), 418–447 (2004)
47. Ph. Clément, S. Li, Abstract parabolic quasilinear equations and application to a groundwater flow problem. Adv. Math. Sci. Appl. **3** (Special Issue), 17–32 (1993/1994)
48. Ph. Clément, J. Prüss, in *An Operator-Valued Transference Principle and Maximal Regularity on Vector-Valued L_p-Spaces*, Evolution Equations and their Applications in Physics and Life Sciences, ed. by G. Lumer, L. Weis (Marcel Dekker, New York, 2000) pp. 67–87
49. T. Coulhon, L. Saloff-Coste, Puissances d'un opérateur régularisant. Ann. Inst. Henri Poincaré, Prob. Stat. **26**, 419–436 (1990)
50. C. Cuevas, Weighted convergent and bounded solutions of Volterra difference systems with infinite delay. J. Differ. Equat. Appl. **6**(4), 461–480 (2000)
51. C. Cuevas, L. del Campo, An asymptotic theory for retarded functional difference equations. Comput. Math. Appl. **49**(5–6), 841–855 (2005)
52. C. Cuevas, J.C. de Souza, A pertubation theory for the discrete harmonic oscillator equation. J. Differ. Equat. Appl. **16**(12), 1413–1428 (2010)
53. C. Cuevas, C. Lizama, Maximal regularity of discrete second order Cauchy problems in Banach spaces. J. Differ. Equat. Appl. **13**(12), 1129–1138. (2007)
54. C. Cuevas, C. Lizama, Semilinear evolution equations on discrete time and maximal regularity. J. Math. Anal. Appl. **361**, 234–245 (2010)
55. C. Cuevas, C. Lizama, Semilinear evolution equations of second order via maximal regularity. Adv. Difference Equat. **2008**, 20 (2008), Article ID 316207
56. C. Cuevas, M. Pinto, Asymptotic behavior in Volterra difference systems with unbounded delay. J. Comput. Appl. Math. **113**(1–2), 217–225 (2000)
57. C. Cuevas, M. Pinto, Convergent solutions of linear functional difference equations in phase space. J. Math. Anal. Appl. **277**(1), 324–341 (2003)
58. C. Cuevas, C. Vidal, Discrete dichotomies and asymptotic behavior for abstract retarded functional difference equations in phase space. J. Differ. Equat. Appl. **8**(7), 603–640 (2002)
59. C. Cuevas, C. Vidal, A note on discrete maximal regularity for functional difference equations with infinite delay. Adv. Differ. Equat. **2006**, 1–11 (2006)
60. C. Cuevas, C. Lizama, H. Henríquez, On the existence of almost automorphic solutions of Volterra difference equations. J. Differ. Equat. Appl. **18**(11), 1931–1946 (2012)
61. J.M. Cushing, *Difference Equations*, In: A. Hastings and L. Gross (Eds.) Sourcebook in Theoretical Ecology (University of California Press, 2012)
62. L. De Simon, Un' applicazione della theoria degli integrati singalari allo studio delle equazioni differenziali lineare abtratte del primo ordine. Rend. Sem. Math., Univ. Padova, 205–223 (1964)
63. L. del Campo, M. Pinto, C. Vidal, Almost and asymptotically almost periodic solutions of abstract retarded functional difference equations in phase space. J. Difference Equat. Appl. **17**(6), 915–934 (2011)
64. D.F. Delchamps, Stabilizing a linear system with quantized state feedback. IEEE Trans. Automat. Control **35** (8), 916–924 (1990)
65. R. Denk, M. Hieber, J. Prüss, $R-$boundedness, Fourier multipliers and problems of elliptic and parabolic type. Mem. Amer. Math. Soc. **166**(788), (2003)
66. R. Denk, T. Krainer, $R-$boundedness, pseudodifferential operators, and maximal regularity for some classes of partial differential operators. Manuscripta Math. **124**, 319–342 (2007)
67. B. De Pagter, W.J. Ricker, $C(K)$-representation and R-boundedness. J. London Math. Soc. **76**(2), 498–512 (2007)
68. G. Dore, L^p *Regularity for Abstract Differential Equations*, Functional Analysis and Related Topics, Lectures Notes Math. vol. 1540, (Springer, New York, 1991) p. 25–38

69. G. Dore, Maximal regularity in L^p spaces for an abstract Cauchy problem Adv. Differ. Equat. **5**(1–3), 293–322 (2000)

70. A. Drozdowicz, On the asymptotic behavior of solutions of the second order difference equations. Glas. Mat. Ser. III **22**(42), 327–333 (1987)

71. A. Drozdowicz, J. Popenda, Asymptotic behavior of the solutions of an n-th order difference equation. Comment. Math. Prace Mat. **29**(2), 161–168 (1990)

72. A. Drozdowicz, J. Popenda, Asymptotic behavior of the solutions of the second order difference equation. Proc. Amer. Math. Soc. **99**(1), 135–140 (1987)

73. N. Dungey, A note on time regularity for discrete time heat kernels. Semigroup Forum **72**, 404–410 (2006)

74. R. Edwards, G. Gaudry, *Littlewood-Paley and Multiplier Theory* (Springer, Berlin, 1977)

75. S. Elaydi, *An Introduction to Difference Equations*, Undergraduate Texts in Mathematics, 3rd edn. (Springer, New York, 2005)

76. S. Elaydi, S. Murakami, E. Kamiyama, Asymptotic equivalence for difference equations with infinite delay. J. Differ. Equat. Appl. **5**(1), 1–23 (1999)

77. K. Engel, R. Nagel, *One-Parameter Semigroups for Linear Evolution Equations* (Springer, New York, 2000)

78. J. Esterle, *Quasimultiplier, Representation of H^∞, and the Closed Ideal Problem for Commutative Banach Algebras*, In radical Banach algebras and automatic continuity (Long Beach, California, 1981), Lecture Notes in Mathematics **975**, (Springer, Berlin, 1983), pp. 66–162

79. S. Fackler, The Kalton-Lancien theorem revisited: Maximal regularity does not extrapolate. J. Funct. Anal. **266**(1), 121–138 (2014)

80. S.R. Foguel, A counterexample to a problem of Sz.-Nagy. Proc. Amer. Math. Soc. **15**, 788–790 (1964)

81. T. Fort, *Finite Difference and Difference Equations in the Real Domain* (Oxford University Press, Oxford, 1948)

82. T. Furumochi, S. Murakami, Y. Nagabuchi, A generalization of Wiener's lemma and its application to Volterra difference equations on Banach spaces. J. Difference Eqs. Appl. **10**(13–15), 1201–1214 (2004)

83. T. Furumochi, S. Murakami, Y. Nagabuchi, *Stabilities in Volterra Difference Equations on a Banach Space*, In Differences and Differential Equations, vol. 42 Fields Institute Communications, American Mathematical Society (Providence, RI, 2004) pp. 159–175

84. T. Furumochi, S. Murakami, Y. Nagabuchi, Volterra difference equations on a Banach space and abstract differential equations with piecewise continuous delays. Japan. J. Math. (N.S.), **30**(2), 387–412 (2004)

85. Z. Gajić, M. Qureshi, *Lyapunov Matrix Equations in System Stability and Control* (Academic Press, New York, 1995)

86. M. Geissert, Maximal L^p regularity for parabolic difference equations. Math. Nach. **279**(16), 1787–1796 (2006)

87. D. Gilbarg, N.S. Trudinger, *Elliptic Partial Differential Equations of Second Order*, Grundlehren der mathematischen Wissenschaften, vol. 224 (Springer, Newyork, 1983)

88. M. Girardi, L. Weis, Operator-valued Fourier multiplier theorems on $L^p(X)$ and geometry of Banach spaces. J. Funct. Anal. **204**(2), 320–354 (2003)

89. S. Goldberg, *Introduction to Difference Equations* (Wiley, New York, 1958)

90. S. Guerre-Delabrière, L_p-regularity of the Cauchy problem and the geometry of Banach spaces. Illinois J. Math. **39**(4), 556–566 (1995)

91. D. Guidetti, S. Piskarev, Stability of the Crank-Nicolson scheme and maximal regularity for parabolic equations in $C^\theta(\overline{\Omega})$ spaces. Numer. Funct. Anal. Optim. **20**(3–4), 251–277 (1999)

92. I. Győri, D. Reynolds, Sharp conditions for boundedness in linear discrete Volterra equations. J. Difference Equat. Appl. **15**(11–12), 1151–1164 (2009)

93. I. Győri, D. Reynolds, On admissibility of the resolvent of discrete Volterra equations. J. Differ. Equat. Appl. **16**(12), 1393–1412 (2010)

94. M. Haase, Y. Tomilov, Domain characterization of certain functions of power-bounded operators. Studia Math. **193**(3), 265–288 (2010)
95. V. Hahn, P. Parks, *Stability Theory* (Prentice-Hall, Englewood Cliffs, NJ, 1993)
96. J. Hale, J. Kato, Phase space for retarded equations with infinite delay. Funkciolaj Ekvacioj **21**, 11–41 (1978)
97. Y. Hamaya, Existence of an almost periodic solution in a difference equation with infinite delay. J. Differ. Equat. Appl. **9**(2), 227–237 (2003)
98. M. Hieber, S. Monniaux, Pseudo-differential operators and maximal regularity results for non-autonomous parabolic equations. Proc. Amer. Math. Soc. **128**(4), 1047–1053 (2000)
99. Y. Hino, S. Murakami, T. Naito, *Functional Differential Equations with Infinite Delay*, Lectures Notes in Mathematics, vol. 1473 (Springer, Berlin, 1991)
100. T. Hytönen, R-boundedness and multiplier theorems. Ph.D. Thesis, Helsinski University of Technology, 2000
101. T. Hytönen, Convolution, multipliers and maximal regularity on vector-valued Hardy spaces. J. Evol. Equat. **5**, 205–225 (2005)
102. V.I. Istrǎţescu, *Fixed Point Theory, An Introduction*, Mathematics and Its Applications, vol. 7 (D. Raider Publishing Company, Dordrech, Holland, 1981)
103. D. Jagermann, *Difference Equations with Applications to Queues*, vol. 233, Pure and Applied Mathematics (Marcel Dekker, New York, 2000)
104. R.C. James, Some self dual properties of normed linear spaces. Ann. Math. Studies **69**, 159–168 (1972)
105. R.C. James, Super-reflexive spaces with bases. Pacific J. Math. **41**, 409–419 (1972)
106. R.C. James, Super-reflexive Banach spaces. Can. J. Math. **24**, 896–904 (1972)
107. W.B. Johnson, J. Lindenstrauss, Handbook of the Geometry of Banach Spaces, vol. 1, ed by W.B. Johnson, J. Lindenstrauss (Elsevier, Amsterdam, 2001)
108. E.I. Jury, *Theory and Applications of the Z-Transform Method*, (Robert E. Kreiger, Florida, 1982)
109. M. Kac, *Statical Independence in Probability, Analysis and Number Theory* (American Mathematical Society, 1959)
110. N.J. Kalton, S. Montgomery-Smith, K. Oleszkiewicz, Y. Tomilov, Power-bounded operators and related norm estimates. J. London Math. Soc. **2**(70), 463–478 (2004)
111. N. J. Kalton, P. Portal, Remarks on l^1 and l^∞ maximal regularity for power bounded operators. J. Aust. Math. Soc. **84**(3), 345–365 (2008)
112. N. J. Kalton, G. Lancien, A solution of the problem of L_p maximal-regularity. Math. Z. **235**, 559–568 (2000)
113. N. J. Kalton, L. W. Weis, The H^∞-calculus and sums of closed operators. Math. Ann. **321**(2), 319–345 (2001)
114. Y. Katznelson, L. Tzafriri, On power-bounded operators. J. Funct. Anal. **68**, 313–328 (1986)
115. S. McKee, J. Popenda, On the existence of asymptotically constant solutions of a system linear difference equations. Fasc. Math. **28**, 109–117 (1998)
116. W.G. Kelley, A.C. Peterson, *Difference Equations*, 2nd edn. (Academic, New York, 1991/2000)
117. V. Keyantuo, C. Lizama, Fourier multipliers and integro-differential equations in Banach spaces. J. London Math. Soc. **69**(3), 737–750 (2004)
118. V. Keyantuo, C. Lizama, Maximal regularity for a class of integro-differential equations with infinite delay in Banach spaces. Studia Math. **168**(1), 25–49 (2005)
119. V. Keyantuo, C. Lizama, Periodic solutions of second order differential equations in Banach spaces. Math. Z. **253**(3), 489–514 (2006)
120. V. Keyantuo, C. Lizama, V. Poblete, Periodic solutions of integro-differential equations in vector-valued function spaces. J. Differ. Equat. **246**(3), 1007–1037 (2009)
121. V. Kocic, G. Ladas, *Global Behavior of Nonlinear Difference Equations of Higher Order with Applications* (Kluwer, Boston, 1993)
122. V. Kolmanovskii, A. Myshkis, Stability in the first approximation of some Volterra difference equations. J. Differ. Equat. Appl. **3**(5–6), 563–569 (1998)

123. V. Kolmanovskii, L. Shaikhet, Some conditions for boundedness of solutions of difference Volterra equations. Appl. Math. Lett. **16**(6), 857–862 (2003)

124. V. Kolmanovskii, A. Myshkis, J. Richard, Estimate of solutions for some Volterra difference equations. Nonlinear Anal. Ser. A: TMA **40**, 1–8 345–363 (2000)

125. V.B. Kolmanovskii, E. Castellanos-Velasco, J.A. Torres-Muñoz, A survey: stability and boundedness of Volterra difference equations. Nonlinear Anal. **53**(7), 861–928 (2003)

126. P.C. Kunstmann, L. Weis, Perturbations theorems for maximal L^p−regularity. Annali della Scuola Normale Superiore di Pisa, Classe di Scienze 4e série, **30**(2), 415–435 (2001)

127. P.C. Kunstmann, L. Weis, *Maximal L^p−Regularity for Parabolic Equations, Fourier Multiplier Theorems and H^∞−Functional Calculus*, Functional Analysis Methods for Evolution Equations, Lectures Notes in Mathematics, vol. 1855 (Springer, Berlin 2004), pp. 65–311

128. O.A. Ladyzenskaya, V.A. Solonnikov, N.N. Ural'tseva, *Linear and Quasilinear Equations of Parabolic Type* (American Mathematical Society Translations Mathematics Monographs, Providence, R.I., 1968)

129. V. Lakshmikantham, D. Trigiante, *Theory of Difference Equations and Applications* (Academic Press, New York, 1988)

130. V. Lakshmikantham, D. Trigiante, *Theory of Difference Equations, Numerical Methods and Applications*, Pure and Applied Mathematics, 2nd edn. (Marcel Dekker, New York, 2002)

131. Y. Latushkin, F. Räbiger, Operator valued Fourier multiplier and stability of strongly continuous semigroup. Integr. Equat. Oper. Theor. **51**(3), 375–394 (2005)

132. C. LeMerdy, Counterexamples on L^p-maximal regularity. Math. Z. **230**, 47–62 (1999)

133. J. Lindenstrauss, L. Tzafriri, *Classical Banach Spaces II* (Springer, Berlin, 1996)

134. C. Lizama, Fourier multiplier and periodic solutions of delay equations in Banach spaces. J. Math. Anal. Appl. **324**, 921–933 (2006)

135. C. Lizama, V. Poblete, Maximal regularity of delay equations in Banach spaces. Studia Math. **175**, 91–102 (2006)

136. W. Long, W-H. Pan, Asymptotically almost periodic solution to a class of Volterra difference equations. Adv. Differ. Equat. **2012**, 199 (2012)

137. H.P. Lotz, Uniform convergence of operator on L^∞ and similar spaces. Math. Z. **190**(2), 207–220 (1985)

138. Y. Lyubich, Spectral localization, power boundedness and invariant subspaces under Ritts type condition. Studia Math. **134**, 153–167 (1999)

139. M. Ma, J. Yu, Existence of multiple positive periodic solutions for nonlinear functional difference equations. J. Math. Anal. Appl. **305**, 483–490 (2005)

140. H. Matsunaga, S. Murakami, Some invariant manifolds for functional difference equations with infinite delay. J. Differ. Equat. Appl. **10**(7), 661–689 (2004)

141. H. Matsunaga, S. Murakami, Asymptotic behavior of solutions of functional difference equations. J. Math. Anal. Appl. **305**(2), 391–410 (2005)

142. H. Matsunaga, S. Murakami, Y. Nagabuchi, Y. Nakano, Formal adjoint equations and asymptotic formula for solutions of Volterra difference equations with infinite delay. J. Difference Equat. Appl. **18**(1), 57–88 (2012)

143. B. Maurey, G. Pisier, Séries de variables aléatoires vectorielles indépendantes et propriétés géométriques des espaces de Banach. Studia Math. **58**, 45–90 (1976)

144. R. Mickens, *Difference Equations* (Van Nostrand, Reinhold, New York, 1990)

145. Sh. E. Mikeladze, De la résolution numérique des equations intégreles. Bull. Acad. Sci. URSS, **VII**, 255–257 (1935)(in Russian).

146. K.S. Miller, *Linear Difference Equations* (W.A. Benjamin, New York, 1968)

147. S. Monniaux, Maximal regularity and applications to PDEs. *Analytical and numerical aspects of partial differential equations*, 247–287, *Walter de Gruyter, Berlin*, (2009)

148. S. Murakami, Representation of solutions of linear functional difference equation in phase space. Nonlinear Anal. T.M.A. **30**(2), 1153–1164 (1997)

149. S. Murakami, Some spectral properties of the solution operator for linear Volterra difference system. In *Proceedings of the Third International Conference On Difference Equations*, Taipe, China, 1997, pp. 301–311

150. S. Murakami, Stabilities with respect to a weight function in Volterra difference equations. Advances in discrete dynamical systems, Math. Soc. Japan, Tokyo, Adv. Stud. Pure Math. **53**, 179–187 (2009)

151. Y. Nagabuchi, Decomposition of phase space for linear Volterra difference equations in a Banach space. Funkcial. Ekvac. **49**(2), 269–290 (2006)

152. B. Nagy, J.A. Zemánek, A resolvent condition implying power boundedness. Studia Math. **134**, 143–151 (1999)

153. G.N. Nair, R.J. Evans, Stabilization with data-rate-limited feeedback: Tightest attainable bound. Systems Control Lett. **41**(1), 49–56 (2000)

154. K. Ogata, *Discrete Time Control Systems* (Printice Hall, New Jersey, 1987)

155. B.G. Pachpatte, *Integral and Finite Difference Inequalities and Applications*, ed. by J. Van Mill. Mathematics Studies, vol. 205 (Elsevier/North Holland, Amsterdam, 2006)

156. C. Palencia, S. Piskarev, On multiplicative perturbations of C_0-groups and C_0-cosine operator families. Semigroup Forum, **63**, 127–152 (2001)

157. V.N. Phat, J. Jiang, Stabilization of nonlinear discrete-time systems via a digital communication channels. Int. J. Math. Sci. **1**, 43–56 (2005)

158. G. Pisier, Probabilistic methods in the geometry of Banach spaces, CIME Summer School 1985, Springer Lectures Notes, **1206**, 167–241 (1986)

159. V. Poblete, Solutions of second-order integro-differential equations on periodic Besov spaces. Proc. Edinb. Math. Soc. **50**, 477–492 (2007)

160. P. Portal, Discrete time analytic semigroups and the geometry of Banach spaces. Semigroup Forum **67**, 125–144 (2003)

161. P. Portal, Analyse harmonique des fonctions à valeurs dans un espace de Banach pour l' étude des equations d'évolutions. Ph.D. Thesis, Université de Franche-Comté, 2004

162. P. Portal, Maximal regularity of evolution equations on discrete time scales. J. Math. Anal. Appl. **304**, 1–12 (2005)

163. R.B. Potts, Differential and difference equations. Am. Math. Monthly **89**, 402–407 (1982)

164. R.K. Ritt, A condition that $\lim_{n \to \infty} n^{-1}T^n = 0$, Proc. Amer. Math. Soc. **4**, 898–899 (1953)

165. J.L. Rubio de Francia, in *Martingale and Integral Transform of Banach Space Valued Function*, ed by J. Bastero, M. San Miguel, Probability and Banach Spaces (Proceeding Zaragoza 1985), Lectures Notes Mathematics, vol. 1221 (Springer, Berlin, 1986), pp. 195–222

166. J. Saal, Maximal regularity for the Stokes system on noncylindrical space-time domains. J. Math. Soc. Japan **58**(3), 617–641 (2006)

167. A. Sasu, Exponential dichotomy and dichotomy radius for difference equations. J. Math. Anal. Appl. **344**, 906–920 (2008)

168. H. Schmeisser, H. Triebel, *Topics in Fourier Analysis and Function Spaces* (Geest and Portig, Leipzig, Wiley, Chichester 1987)

169. P.E. Sobolevskii, Coerciveness inequalities for abstract parabolic equations. Soviet Math. (Doklady), **5**, 894–897 (1964)

170. Y. Song, Almost periodic solutions of discrete Volterra equations. J. Math. Anal. Appl. **314**(1), 174–194 (2006)

171. Y. Song, Periodic and almost periodic solutions of functional difference equations with finite delay. Adv. Differ. Equat. **2007**, 15 (2007), Article ID 68023

172. Y. Song, Asymptotically almost periodic solutions of nonlinear Volterra difference equations with unbounded delay. J. Differ. Equat. Appl. **14**(9), 971–986 (2008)

173. Y. Song, H. Tian, Periodic and almost periodic solutions of nonlinear discrete Volterra equations with unbounded delay. J. Comput. Appl. Math. **205**(2), 859–870 (2007)

174. Z. Štrkalj, L. Weis, On operator-valued Fourier multiplier theorems. Trans. Amer. Math. Soc. **359**(8), 3529–3547 (2007)

175. H. Triebel, *Theory of Function Spaces*, Monographs in Mathematics, vol. 78 (Birkhäuser, Basel, 1983)

176. M.C. Veraar, L.W. Weis, On semi-R-boundedness and its applications. J. Math. Anal. Appl. **363**, 431–443 (2010)

177. J. van Nerven, M.C. Veraar, L.W. Weis, Stochastic evolution equations in UMD Banach spaces. J. Funct. Anal. **255**(4), 940–993 (2008)

178. C. Vidal, Existence of periodic and almost periodic solutions of abstract retarded functional difference equations in phase spaces. Adv. Differ. Equat. **2009**, 19 (2009) Article ID 380568

179. L. Weis, *A New Approach to Maximal L_p-Regularity*, Lecture notes Pure Applied mathematics, vol. 215 (Marcel Dekker, New York, 2001) pp. 195–214

180. L. Weis, Operator-valued Fourier multiplier theorems and maximal L_p-regularity. Math. Ann. **319**, 735–758 (2001)

181. H. Witvliet, Unconditional Schauder decomposition and multiplier theorems. Ph.D. Thesis, Techniche Universitet Delft, 2000

182. F. Zimmermann, On vector-valued Fourier multiplier theorems Studia Math. **93**, 201–222 (1989)

Index

R.P. Agarwal et al., *Regularity of Difference Equations on Banach Spaces*,
DOI 10.1007/978-3-319-06447-5, © Springer International Publishing Switzerland 2014

Printed in the United States
By Bookmasters